ISSUES IN THEORETICAL DIVERSITY

PHILOSOPHICAL STUDIES SERIES

VOLUME 106

Founded by Wilfrid S. Sellars and Keith Lehrer

The titles published in this series are listed at the end of this volume.

ISSUES IN THEORETICAL DIVERSITY

Persistence, Composition, and Time

by

KRISTIE LYN MILLER

The University of Sydney, Sydney, Australia

Springer

A C.I.P. Catalogue record for this book is available from the Library of Congress.

ISBN-10 1-4020-5255-3 (HB)
ISBN-13 978-1-4020-5255-2 (HB)
ISBN-10 1-4020-5256-1 (e-book)
ISBN-13 978-1-4020-5256-9 (e-book)

Published by Springer,
P.O. Box 17, 3300 AA Dordrecht, The Netherlands.

www.springer.com

Printed on acid-free paper

Contents

Introduction

This is a book about objects: about their persistence, and their composition. It's about under what circumstances smaller things compose larger things: is it the case that wherever we have some objects, there is some further object that is composed of those objects, or do objects only compose some further object under certain conditions? It's about whether objects are ultimately composed of simples—objects that lack any parts—and if so, what are those simples like? It's about the nature of the composition relation itself: is mereology—the theory of parts and wholes—the only account of composition, or might there be some non-mereological account of the relation between objects across time? It's about the manner in which objects are related to one another. Are there distinct objects that share all of the same matter at certain times, and if so, how are they related? It's about how objects exist through time: how do objects both change across time, and yet remain the same object? Are persisting objects three-dimensional things that wholly exist at each moment, or are they temporally extended four-dimensional things that are only partly present at each moment?

But this is also a book about theories and theoretical diversity. It's about what it might mean to say that two theories are in some sense equivalent, and about how we could ever know that this is the case. Why combine the first-order metaphysics of objects, with the second-order metaphysics of theoretical equivalence? Because, as I explain shortly, these two projects are deeply inter-related. We can only fully understand an account of theoretical equivalence in the context of a live debate about the status of particular theories. And we can only answer many first-order metaphysical questions, once we have settled the second-order debate. So these two tasks must be approached together.

What then, do I mean when I talk of theoretical equivalence? Well, it is a common idea that there is a sense in which two theories might amount to the same thing, or that where we think we have distinct theories, in fact we have a case of mere verbal disagreement. The task of constructing an account of equivalence is most intractable within the domain of metaphysics where, perhaps unlike in the sciences, it seems plausible that mere empirical equivalence is not sufficient for theoretical equivalence. Yet within metaphysics we often find claims being made about the equivalence of theories. We find such claims with respect to (putatively) different theories of time—presentism and eternalism—with respect to (putatively) different theories of composition—restricted composition and unrestricted composition—and with respect to (putatively) different theories of ontology—realism and fictionalism. To make any sense of these claims we need some account of the equivalence relation: we need to know what it would take for theories to be equivalent.

This book develops an account of a strong kind of theoretical equivalence: an equivalence that holds between theories that, intuitively speaking, are simply different ways of describing the same underlying reality. I call this type of equivalence *metaphysical equivalence*, and in chapter one I develop an account of metaphysical equivalence in terms of theories being correctly inter-translatable. To address the issues of how we could ever know that theories are equivalent in this manner, I formulate a set of *diagnostic criteria* against which we can measure theories to determine whether or not they are equivalent.

I then explore the issue of theoretical diversity by considering, in the light of this framework, two metaphysical theories that are widely held to be rivals. For the best way to flesh out an account of equivalence is to bring its resources to bear on theories for which it is controversial whether an equivalence relation holds. And it is controversial indeed whether the theories of three- and four-dimensionalism—two apparently rival theories of persistence—might be equivalent. So it is no accident that I choose these two theories to consider. But that they are controversial candidates for being equivalent is not all that makes them a good choice. For, as I will argue, answering the question of how objects persist through time—and hence whether three- and four-dimensionalism are equivalent—involves resolving a gamut of other key issues in metaphysics: issues regarding the nature of composition, the nature of simples, and ultimately, the nature of objects themselves.

Thus in the following chapters I argue that it is a mistake to see three-dimensionalism and four-dimensionalism as single monolithic theories that stand in opposition to one another. In chapter four I disentangle a number of different versions of three- and four-dimensionalism. Some of this work involves showing that what has hitherto been treated as a single theory, are in fact multiple theories over which there has been equivocation. Some of the work involves developing wholly novel theories of persistence that fall under the broad umbrella terms of 'four-dimensionalism' and 'three-dimensionalism'. In some sense, then, I argue that there is far greater theoretical diversity than has previously been thought.

Yet ultimately I argue that three- and four-dimensionalism are metaphysically equivalent. How can both of these claims be true? It is because talk of a theory of persistence is ambiguous between, on the one hand, talk of a complete theory of an object's persistence through time, and on the other hand, talk of a theory regarding what we might call the *dimension of persistence*. When we talk about a complete theory of persistence, we are talking about a theory that includes a whole range of metaphysical commitments: commitments to an account of time, an account of composition, and an account of the particulars from which composites are composed. In this sense, a theory of persistence is a metaphysical *package* that includes a number of components: it is a package that tells us what it is, and what it takes, to be an object at, and across time. One of the components that such a package includes, is an account of the dimension of persistence. This is the second sense in which one might talk of a theory of persistence. An account of the dimension of persistence is simply an account of whether persisting objects are four-dimensional or three-dimensional. It is in the former sense that I develop a number of novel theories of persistence.

What is important about these theories is that they offer the possibility of a genuinely new way of thinking about objects and their persistence, and hence offer a new approach to some of the intractable problems of persistence. Moreover, in considering these theories we develop a better understanding of which components of the theory—which bits of the package, if you will—are playing what role in the overall account. It might, for instance, turn out that whichever theory one adopts, some components of the package remain unchanged. If, along the dimension of persistence, three-dimensionalism turned out to be false, or incoherent, then all of the packages would be the same along the dimension of persistence: they would all be four-dimensionalist. But various other components might be different. So there could still be very different four-dimensionalist theories of persistence. Indeed, I argue that something like this is true. It is not that there is one component of the packages that is invariant across all of the complete theories, rather, it is the case that the most successful theories of persistence are those packages that all share a particular component. And that component is not a particular account of the dimension of persistence.

Indeed, I argue quite the reverse with respect to the dimension of persistence. Bringing to bear the resources of the diagnostic criteria of metaphysical equivalence, I argue that for any three-dimensionalist package, there is an *analogous* four-dimensionalist package that is metaphysically equivalent and *vice versa*. In this sense pairs of theories are analogous just if they differ with respect to only one component of the package: the dimension of persistence. Then if such theories are equivalent, then there is equivalence across the dimension of persistence. But until we develop each of these complete theories, it is not easy to see this: for we tend to compare one metaphysical package that contains three-dimensionalism as one of its components, with a substantially different metaphysical package that contains four-dimensionalism as one of its components.

So developing each of these theories not only allows us to evaluate novel accounts of persistence, and in particular, the role of each of the components of that theory, but also allows us to evaluate a particular claim about theoretical diversity. But it allows us to do more than that. Developing each of these theories involves, in each case, developing a conception of what it is to be a composite object both at, and over time. These different theories have very different answers to the question of what it is to be a composite object, and what it is to be a *persisting* composite object. Examining each of them provides an impressive view of the logical terrain of the ontology of objects. It tells us how certain views about the nature of composite objects, entail certain views about the nature of simples. It tells us how certain views about the nature of persistence, entail certain views about the nature of composition. It tells us what sorts of views about simples, composition and persistence go together consistently in a metaphysical package. In considering these complete theories of persistence, I argue that we make startling discoveries about the interrelations between the various components of the packages. We discover that adopting certain views as a component of a theory, entails adopting other components, and we find that frequently it was not at all obvious that this was the case. Ultimately then, considering these theories tells us not only about theoretical diversity, but crucially, it also tells us about the structure of theoretical space when it comes to the ontology of objects.

The successful completion of this book is due to the hard work and forbearance of a number of persons. First there is my poor Pomeranian, usually known as the furry ball, who has forfeited countless walks and sat on any number of hastily drawn diagrams of space-time worms. Many chapters have been typed over the top of her head as she silently muses on the fact that someone could worry about unrestricted composition when it's just plain clear that there could be nothing composed of a dog and a cat.

Thanks also go to a number of people who have read and commented on various chapters. They include Mark Colyvan and Dominic Hyde. Thanks also go to Jay Garfield for sage advice along the way. The biggest thanks of all go to David Braddon-Mitchell, whose tireless support has never flagged, even at the worst of moments. He is the Anselmian friend—one than which no greater can be conceived. There is no problem I have encountered in this book to which he has not listened and provided help. There is no chapter that he has not read and commented on. There is no idea I have had that we have not discussed. That the book exists at all is testament to his unwavering patience. If it is ever said in the future, that I have contributed anything in any way to the world, it will be because of his support. He will always have my gratitude and my friendship.

Finally, thanks go to a number of publishers who have allowed material that has appeared elsewhere to be re-printed here. Chapter one can be found in slightly modified form as 'What is Metaphysical Equivalence?' (2005). *Philosophical Papers* 34(1) 35–74. Chapter three owes its origins to two papers: 'A New Definition of Endurance' *Theoria* (4): 209–332, and to 'There is no *simpliciter simpliciter*', (with D. Braddon-Mitchell) *Philosophical Studies* forthcoming. Chapter four owes its Origins to 'Blocking the path from vagueness to four-dimensionalism' (2005) in *Ratio* 18(3) 317–331, and also from 'Non-mereological Universalism' forthcoming in *European Journal of Philosophy*. Chapters five and six owe their origins to 'The Metaphysical Equivalence of Three and Four Dimensionalism' *Erkenntnis* 62(1) 91–117. Chapter seven owes its origin to 'Travelling in Time: How to wholly exist in Multiple Places at the Same Time' forthcoming in *Canadian Journal of Philosophy*. Some of chapter five owes its origins to 'Sparse Parts' *Sorites* 17 (1) in print. Chapter eight owes its origins to 'Enduring Special Relativity' (2004). *Southern Journal of Philosophy* 42(3): 349–370.

Chapter 1

WHAT IS METAPHYSICAL EQUIVALENCE?

1. METAPHYSICAL EQUIVALENCE

In any domain of enquiry it is frequently the case that we have what appear to be rival theories: theories that make different claims about the way the world is, or perhaps, how the world must be. Debate between defenders of rival theories then takes a standard form, with each side providing arguments of one sort or another in support of their preferred view, and launching objections against the alternative view. Sometimes, however, there is in addition to this standard debate, a meta-debate about whether the theories in question are *really* rivals at all, or whether they are in some sense equivalent theories. Those who think that putatively rival theories are not rivals at all, think that there is no substance to the standard debate: debaters are either simply talking past one another, or at best, are providing purely pragmatic reasons to prefer one theory over another.

It is not always clear, though, exactly what is meant, on various occasions, by the claim that two theories are really equivalent. In this chapter I try to provide a clearer account of what we might mean by such a claim. To get a feel for the notion of theoretical equivalence, I begin by briefly describing a number of areas in which something like this claim is made about apparently rival theories. Consideration of these cases will provide a flavour for the general notion of theoretical equivalence, and they will later serve as case studies as I go on to develop a more precise definition of metaphysical equivalence. Then with a definition in mind, the question becomes how we are to know, on any occasion, whether two or more theories are equivalent. To address this issue I develop what I call *diagnostic criteria* of metaphysical equivalence. These are criteria against which putatively equivalent theories can be measured, and it can be determined whether they are equivalent or not.

First two issues. This chapter does not attempt to provide a formal account of equivalence. There are certainly advantages to be gained by developing a formal account: rigour being prime among them. In practical terms though, a formal account has its limitations. The sorts of complex metaphysical theories about which equivalence claims are made, are theories that could be formalised in a number of ways. Then whether putatively rival theories turn out to be equivalent or not, will depend on which way we choose to formalise each theory. Given one way of formalising each theory, we may conclude that they are not equivalent, and given another way of formalising each theory, we may conclude that they are equivalent. Since a purely formal account of equivalence provides us with no guidance regarding which of the various ways of formalising each theory is the correct one, in practical terms such an account may often offer us no way to determine whether theories are equivalent or not. And this should hardly be surprising. After all, we are considering theories that at least *appear* to be rivals. If in fact they turn out to be equivalent, then plausibly (particularly in the case of metaphysical theories) this is because what appeared to be shared terminology was actually not shared at all: the same term turned out to have a different meaning in each theory. What explained why they appeared to be rival theories was that it seemed as though there were claims of one theory that straightforwardly contradicted claims of the other theory: one theory claimed that *P*, and another claimed that not *P*. Such claims turn out not to be contradictory only if '*P*', in one theory, means something different to '*P*' in the other theory. But if it is in virtue of features like this, that apparently rival theories turn out to be equivalent, then we can only expect that a formal account will have limitations. For formalising each theory precisely requires us to determine how we will understand terms like '*P*' in each theory, and nothing about the formalism tells us that.

So the account of equivalence that I give is a looser account, and I make no apologies for that. The hope is that a looser account, though less rigorous, will be one that provides some tools to help us in determining whether the sorts of real life metaphysical theories that are the subject of claims about equivalence, really are equivalent or not. Through the remainder of the book we see this account at work, and what we discover is, I think, instructive.

The second issue to note is that in examining the various case studies and developing diagnostic criteria in part based on such an examination, I do not presuppose that any or all of the cases are ones in which we have genuine metaphysical equivalence. So I do not attempt to argue that the claims of equivalence in these cases are either true or false; rather, I consider how the world would need to be if the claims were true, and how we could come to know that the claims were true if indeed they were.

1.1 Case Studies

It is frequently argued, or at least hinted, that certain theories are in *some* way equivalent. Presentists are sometimes accused of espousing a metaphysics that is either trivially false, or equivalent to eternalism. It is suggested that although it appears that presentists and eternalists are making substantially different claims about what exists, specifically about whether any temporal locations other than the present exist, this might be mere verbal disagreement. Perhaps the presentist and the eternalist mean something different by 'everything' so that when the presentist says that everything that exists, exists in the present, and the eternalist denies this, they are not making contradictory claims at all.[1]

Equally, there are those who wonder whether what appear to be incompatible ontological claims are really incompatible. Putnam asks us to imagine a world in which there exist three simples. The mereological universalist, who holds that for any arbitrary set of concrete particulars there exists a fusion of the members of that set, (the view that any combination of simples composes some object) will conclude that there exist a total of seven objects in that world. The mereological or compositional nihilist, who holds that there do not exist any composite objects (no combination of simples ever composes some object), will hold that there exist a total of three objects in that world. Putnam famously argues that these two views are equivalent. In terms of the universalist language, it is true that there exist seven objects, and false that there exist only three objects, and the reverse is true in terms of the nihilist language. Each of these languages are equally good ways of talking about the world, and since there is no 'absolute' framework from which we can talk about the world as it is in itself, it makes no sense to try to say that there are 'really' seven objects not three, or three objects not seven.[2]

Perhaps these theories are equivalent because 'object' is defined by its role in the entire ontological theory in which it features, and thus universalists, nihilists and non-universalists simply mean something different when they deny or affirm that certain arrangements of simples compose some object.[3] Or perhaps, as has recently been argued by Eli Hirsch, the existential quantifier can have multiple meanings. On some interpretations of 'there exists a thing' certain sentences will be true, and on other interpretations they will be false,[4] thus explaining how universalists, non-universalists and nihilists can all speak truly.

Alan Sidelle goes further. He argues that in matters of persistence and ontology there are various 'packages' of views each of which preserves a different set of folk intuitions and theoretical ideals. According to Sidelle, there is no fact of the matter which of these packages truly describes the world. Rather, they are merely different ways of making coherent our various intuitive judgements and theoretical ideals.[5]

Finally, in *Platonism and Anti-Platonism in Mathematics*, Mark Balaguer controversially argues that there is only one viable version of Platonism: full-blooded Platonism, and one viable version of non-Platonism: fictionalism, and that there is a sense in which these two theories are equivalent.[6] Balaguer argues that for all practical purposes there is no difference between Platonism and fictionalism: both offer exactly the same vision of mathematical practice. The only difference between the two theories is with respect to ontology: Platonists maintain that mathematical objects exist, and fictionalists maintain that they do not. Balaguer argues that this is no real difference at all, for there is simply no fact of the matter as to whether mathematical objects exist or not.[7] As Balaguer puts it 'the metaphysical question of whether there exist any abstract objects is empty, but the two conclusions [Platonism and fictionalism] cash this out in different ways.'[8] If Platonism and fictionalism provide the same account of mathematical practice and differ only in matters of ontology, then if the ontological debate is empty it would seem that Platonism and fictionalism are equivalent.

1.2 Defining Metaphysical Equivalence

In all of the examples just mentioned we find claims about theories being in some sense equivalent. In each case the underlying intuition is that theories are equivalent in this sense if somehow they are describing the same underlying reality, albeit using a different language. Thus we find the idea that in these cases we have 'mere verbal disagreement', and that in fact the features of the world described by one theory are identical to the features of the world described by the other theory, and thus there are no facts that could render one theory true and the other false.

This suggests that we might define metaphysical equivalence in terms of sets of worlds being identical. We might say that any two theories x and y are metaphysically equivalent iff the set of the worlds in which x is true, is identical to the set of worlds in which y is true, and the set of worlds in which x is false, is identical to the set of worlds in which y is false. But this will not do. For then any two necessarily true theories will turn out to be metaphysically equivalent.

Ultimately, to deal with theories about necessarily existing abstracta or concreta, we need to appeal to truth makers (choose your favourite account of truth makers). Then a theory about God and atheory about the number 3 are not equivalent, and they are not equivalent because they have different truth makers. At their heart, claims about metaphysical equivalence are claims about truth makers: *any two theories are metaphysically equivalent just in case they have the same truth makers.*

Having defined metaphysical equivalence thus, however, may not seem to be very revealing. Knowing that two theories are equivalent just if they have the same truth makers does not help in determining whether any two particular theories are equivalent or not. But we should not expect a definition to do that job for us. What we need is some apparatus that can be used either to argue that certain theories are metaphysically equivalent, or to dispute such a claim. What we need are diagnostic criteria, and it is to formulating such criteria that I turn in section three. First however, we need to be clear about the relation between metaphysical equivalence and inter-translatability. Are all and only the theories that are metaphysically equivalent correctly inter-translatable? Is correct translatability a criterion of equivalence, does it entail equivalence, or is it just another way of saying that theories are equivalent?

2. INTER-TRANSLATABILITY AND EQUIVALENCE

What is the relation between metaphysical equivalence and the inter-translatability of theories? Putnam holds that any two theories[9] that are correctly inter-translatable are metaphysically equivalent. He argues that universalism—the view that for any two or more objects, there is some further object that is composed of those objects—and nihilism—the view that there exist no composite objects—are inter-translatable, and hence that they are equivalent. The thesis of quantifier variance—the claim that the existential quantifier has, or could have different meanings—goes some way towards explaining how this translation could work. Since we define the logical constants by describing their roles in determining the truth conditions of compound sentences, we can define different meanings of the existential quantifier by stipulating different truth conditions for sentences containing the quantifier. For instance, Hirsch argues that the meaning of the quantifier employed by the universalist is that sentences of the form 'there exists something composed of the *F*-thing and the *G*-thing' are true just in case 'the *F*-thing' refers to something and 'the *G*-thing' refers to something.

The non-universalist's sense of the quantifier—where non-universalists hold that two or more objects compose some further object only under certain conditions—is that sentences of the form 'there exists something composed of the F-thing and the G-thing' are true just if 'the F-thing' and 'the G-thing' refer to things that are connected in certain special ways.[10] While for the nihilist a sentence of the form just described will never be true, since no F-thing or G-thing ever compose anything.

We can see the same notion of translatability arising in Balaguer's arguments. Zalta and Colyvan suggest that one way to understand Balaguer's claim that Platonism and fictionalism are equivalent with respect to all matters but ontology, is to understand Platonism and fictionalism as two interpretations of a single formalism: "∃xAx".[11] Then the disagreement between the two resides in the fact that the Platonist reads the formalism such that the quantifier has existential import, and reads the predicate "A" as "abstract", while the fictionalist reads the quantifier as lacking existential import, and reads the predicate "A" as "fictional". Hence "∃xAx" under one interpretation reads, "there exist abstract objects", and under the other interpretation reads, "there are fictions". Since the debate about whether there exist abstract objects or not is, according to Balaguer, an empty one, we can see these two interpretations as two ways of explicating this ontological emptiness, thus explaining how Platonism and fictionalism provide the same understanding of mathematical practice and yet appear to differ so radically on ontological matters.

The question is whether Putnam is right to hold that if two theories are inter-translatable, this is sufficient for concluding that they are metaphysically equivalent. There seems to be a sense in which Putnam is right. Consider the intuitive sense of a 'correct' translation, according to which a translation is correct just if it truly 'gets it right'. In this intuitive sense, it is difficult to see how a correct translation between theories could fail to be sufficient for metaphysical equivalence. For if the translation is truly correct in this strong sense, then surely the two theories are indeed describing the same underlying reality in different terminology—after all that is, we might think, precisely what it is to have a correct translation.

This claim is strongly supported by the fact that those who reject some particular claim of metaphysical equivalence invariably argue that the theories in question are not correctly inter-translatable, rather than arguing that despite the fact that they are translatable, this is insufficient grounds to conclude that they are equivalent. For instance, Ted Sider, Peter van Inwagen and Trenton Merricks all resist the idea that the existential quantifier has multiple interpretations, and thus resist the claim that nihilism, non-universalism and universalism are equivalent.[12] And presumably fictionalists and Platonists who hold that their theories are not equivalent, think that this is so because they think that the Platonist's sentences that quantify over abstract objects cannot *correctly* by mapped to the fictionalist's sentences that do not so quantify.

So I think we can say that if two theories are correctly inter-translatable, then this entails that they are metaphysically equivalent. What it is for there to exist a correct translation between theories just is for those theories to be describing the same underlying reality: for them to have the same truth makers.

This leaves it open that we might approach the task of determining whether two theories are equivalent or not, by determining whether they are correctly inter-translatable. The real question is whether we can define some function that maps the sentences of one theory onto the sentences of the other theory such that that mapping counts as a correct mapping, and thus counts as a correct translation between those theories.[13] Further, given that we *can* define such a function, what epistemic access can we have to whether or not on any particular occasion, a translation is a correct one. Let us begin with the first question first.

A translation function is a function that maps sentences of one theory onto sentences of some other theory. One sort of translation function is a function that maps the sentences of one theory onto the sentences of another theory just when those sentences are assertible under the same possible situations. Let us call this an *assertibility mapping*. Presumably everyone can agree that, for instance, where the four-dimensionalist will utter 'there is a rabbit temporal part', under the same conditions the three-dimensionalist will utter 'there is a wholly present rabbit'. So too under the same conditions the nihilist will utter 'there exists a dog-wise arrangement of simples', the universalist (and most non-universalists) will utter 'there exists a dog.' In each of these cases we have an assertibility mapping: a function that maps the sentences of theory *A* that are, by the lights of theory *A* correctly assertible, onto the sentences of theory *B* that are, by the lights of theory *B* correctly assertible, when and only when those sentences are assertible under the same possible situations.

Of course, the existence of an assertibility mapping does not show that the theories in question are correctly inter-translatable. Nihilists think that universalists and non-universalists are wrong when they assert that 'there exists a dog' just as fictionalists think that Platonists are wrong when they assert that there exist abstract objects. But any two theories can only be metaphysically equivalent if the sentences they assert under the same circumstances have the same truth-values: if the assertibility mapping is truth preserving. This suggests that we say that an assertibility mapping is a correct mapping—a correct translation—only if it preserves the truth-values of the sentences in each of the theoretical languages. Now *prima facie* we might think that an assertibility mapping that preserves truth values ought to count as a correct translation. If the nihilist and the universalist assert respectively 'there is a dog-wise arrangement of simples' and 'there is a dog' on all and only the same actual and possible occasions, and if on those occasions either both sentences are true, or both false, then don't we have a correct translation between those sentences?

Perhaps we do. But if we have a correct translation just when we have a case of metaphysical equivalence, then it cannot be that an assertibility mapping is a correct mapping iff it preserves truth. For suppose that God necessarily exists. Suppose further that whenever you assert '*x*' I assert '*x* and God exists'. (Or if you are sure that God is contingent, suppose when you assert '*x*' I assert '*x* and p or not p'.) Then it would seem that there is an assertibility mapping between our sentences that is truth preserving. But surely we would be wrong to conclude that '*x*' in your mouth, is a correct translation of '*x* and God exists' in my mouth.

The problem is that we not only want the assertibility mapping to be truth preserving, we want it to be truth preserving in virtue of the same truth makers. That is why a genuinely correct translation between theories entails that they are equivalent. So let us say that we have a *correct translation* between theories just if there is an assertibility mapping that is truth preserving and where it preserves truth in virtue of the same truth makers. Now we are again faced with the problem that since frequently we have no access to truth makers, we do not always have access to whether some assertibility mapping is a correct translation. Thus in these cases we will have no access to whether or not the theories in question are equivalent.

What we need are tools to help us in deciding whether on any occasion we have a correct translation. Let us begin by defining what I will call a *practical translation,* where a practical translation is an assertibility mapping that is truth preserving. Since being a correct translation entails being a practical translation, if we can show that some mapping is not a practical translation, then we can conclude that it is not a correct translation, and therefore that the theories in question are not equivalent. That is, we can at least go some way towards showing how we would *falsify* a claim about metaphysical equivalence. Of course, that we have a practical translation does not entail that we have a correct translation. In sections 3.4 and 4 we will consider what, if anything, licences the move from holding that a translation is practical, to that it is correct. First however, let us see how we can determine whether a translation is a practical one.

3. DIAGNOSTIC CRITERIA

3.1 Diagnosing a Practical Translation: Empirical Equivalence

How are we to know whether or not we have a practical translation? Well, we have a practical translation just if we have an assertibility mapping that is truth preserving. So one way of diagnosing the existence of a practical translation is via empirical equivalence. If we had defined an assertibility mapping as a function that maps sentences of different theories that are assertible under all and only the same *actual* situations, then an assertibility mapping would entail a weak empirical equivalence. That is, it would entail that the theories in question make all of the same observational predictions in the actual world. In fact we defined an assertibility mapping as a function that maps sentences of different theories that are assertible under all and only the same *possible* situations. Thus we have an assertibility mapping only if the theories in question are strongly empirically equivalent, that is, only if they make the same observational predictions in all worlds—there is no actual or possible piece of evidence that could render one theory true and the other false.[14]

So we can use (strong) empirical equivalence as a tool in helping to determine whether we have a practical translation. For if we can show that the theories in question are not empirically equivalent, then we have shown that we do not have an assertibility mapping, and thus we certainly do not have a correct translation. Now, in many cases it might seem obvious that certain theories are empirically equivalent. It seems to be agreed that nihilism and universalism are empirically equivalent, as are Platonism and fictionalism. So too many would argue, are three- and four-dimensionalism. But this latter is a source of debate. There are those who hold that the empirical discoveries of special relativity show that three-dimensionalism is incoherent,[15] or at least very implausible. I will discuss these issues further in chapter eight. For now we need only note that empirical equivalence provides a place to begin in determining whether theories can be correctly translated. Or, I should say, showing that theories *fail* to be empirically equivalent entails that they are not practically translatable and hence not correctly inter-translatable, so determining that theories fail to be empirically equivalent is one way of falsifying a claim of equivalence.

3.2 Diagnosing a Practical Translation: the Principle of Charity

As we have seen, if we can show that two theories are not empirically equivalent, then we have shown that there is no assertibility mapping, and thus no practical translation. Even if we have an assertibility mapping, however, this does not entail that we have a practical translation. For the assertibility mapping might fail to be truth preserving. It is here that we discover a problem. After all, one who holds that two theories *x* and *y* are not equivalent, will surely maintain that of any pair of co-assertible sentences of those theories, at most only one of those sentences is ever true. So consider the following mapping. We map the universalist' sentence (1) 'there is a dog' to the nihilist's sentence (2) 'there is a dog-wise arrangement of simples'. This is an assertibility mapping, since presumably the former utters sentence (1) just when the latter utters sentence (2). But is this mapping truth preserving? Since the nihilist maintains that there aren't any dogs, presumably she will hold that sentence (1) 'there is a dog' is false, while sentence (2) is true. But then the nihilist can argue that the assertibility mapping cannot be truth preserving. And how is the proponent of equivalence in this case to argue that this is not so, beyond appealing to the equivalence of the theories in question?

Suppose we are considering the sentence 'there exists something composed of my dog and your shoe'—call it a Doe. Then the universalist affirms the sentence 'there exists a Doe'. The non-universalist (let us suppose) denies this. She might assert the following 'there exists a dog and a shoe arranged Doe-wise' or perhaps just 'there exists a dog and a shoe'. If we were to map either of these non-universalist sentences onto the universalist sentence, we would have an assertibility mapping. But, as in the case above, it seems that each of the parties might protest that the mapping is not truth preserving. The non-universalist, for instance, will claim that the sentence 'there exists a Doe' is false, while the sentence 'there exists a dog and a shoe' is not false. Hence the mapping is not truth preserving.

But on what basis does the non-universalist conclude that the sentence 'there exists a Doe' is false? 'There exists a Doe' is false when uttered by the non-universalist, given what she means by that sentence. But the non-universalist can only conclude that the universalist's claim is false if she has some theory about what the universalist means when *she* utters that sentence. Suppose that Hirsch is right and there are two *possible* meanings of the existential quantifier and thus two possible languages corresponding to these two different meanings (whether or not these two languages are in fact the languages of the universalist and the non-universalist).

Given the way Hirsch defines the two meanings of the quantifier, it follows that the two languages are indeed practically inter-translatable. It is the case that we have an assertibility mapping, and that this mapping is truth preserving. For any universalist sentence there is some non-universalist sentence that has the same truth conditions in the sense that relative to any context of utterance, both sentences hold true in all and only the same possible situations. But why think that either of those languages are the languages of the universalist or non-universalist? One good reason is surely that if I am universalist attempting to translate the words of my friend the non-universalist, I should expect most of the sentences that she asserts to be true. Some sort of Davidsonian charity tells me that.[16] Or at least, I should expect most of the sentences she asserts not to be *inexplicably* false. The principle of humanity tells me that.

But if I interpret my non-universalist friend as meaning the same as I do by all of her statements of the form 'there exists an *x*', then I must conclude that a great deal of what she says is false, as for instance, her claim that 'there exists no Doe'. Yet that my friend consistently utters falsehoods seems to be quite inexplicable: she is not suffering hallucinations, she has not been tricked by evil demons or anything of that nature. There are no facts such that, once those facts are made clear to her, she would recognise that some of her utterances are false by her own lights. So there seems no good reason to suppose that most of her utterances are false. A more plausible interpretation is that she simply means something different by some of the terms she uses—something along the lines of those defined by Hirsch. And this can hardly be surprising given that theoretical terms are in part defined by the roles they play in the theory in which they are embedded.[17]

None of this implies anything about the relation between ordinary English and either of the sub-languages of the theories in question. One way to make sense of the universalist/non-universalist case is that the English sense of 'there exists' is sufficiently semantically vague that both the universalist and non-universalist meanings of the quantifier are precisifications of the English sense. Hirsch demurs, holding that only the non-universalist sense of the quantifier captures anything of the ordinary English meaning of 'there exists'. The universalist is speaking a language all of her own, and her claim that 'there exists a Doe' is just false in English. But these considerations, while interesting, are not relevant to the issue of whether or not the theories in question are in fact practically or correctly inter-translatable.

If Hirsch is right and the existential quantifier does have two meanings, then so far we have good reason to think that universalism, non-universalism and nihilism are practically inter-translatable. Hirsch may be wrong. There may be a number of reasons why quantifier variance is false. Perhaps non-universalism, universalism and nihilism are not languages that interpret the quantifier differently, but rather, place restrictions on a quantifier with a univocal meaning, and perhaps various considerations pertaining to the alleged incoherence of the vagueness of existence tell us that.[18] But that is not to be discussed here. What matters is that considerations of charity and humanity mean that if we have an assertibility mapping, we have reason to think that that mapping is truth preserving where failing to do so would result in inexplicably interpreting a large proportion of the sentences of one of the theories as false.

3.3 Diagnosing a Practical Translation: Explanatory Power

We might hope, however, that there are some additional tools that will help us to discern whether, on some occasion, we have a practical translation. Considerations of empirical equivalence (and co-assertibility) will not help when we are considering theories such as Platonism and fictionalism. Might considerations of theoretical virtue aid us in determining whether some assertibility mapping is a practical translation? Consider first Alan Sidelle's striking claim that a number of 'ontological packages' are, in some sense, equivalent, insofar as there is no fact of the matter as to which of the packages is the correct one. It is not clear that Sidelle means to claim that the packages are metaphysically equivalent in the sense I have described, since this would require that each of the 'elements' of each package be inter-translatable, and this is a tall order when packages include different components regarding persistence, time and composition.

Sidelle acknowledges that these packages preserve different folk intuitions and have different explanatory virtues. So what is interesting here, is whether *if* the packages were metaphysically equivalent, they would also be explanatorily equivalent. That is, is the fact that they do not appear to be explanatorily equivalent, a reason to conclude that there exists no practical translation? Or is it a reason to conclude that although perhaps there exists a practical translation, there exists no correct translation? If either were the case, then consideration of the explanatory virtues of putatively equivalent theories could help us to determine whether those theories are equivalent.

To address this question, consider the following case. Consider the theories of classical thermodynamics and statistical mechanics, and the theory of meteorology, and some physical theory that describes weather conditions at the microphysical level. These are not the sorts of theories with respect to which we usually tend to worry about metaphysical equivalence, because they are not theories that occur 'at the same level'. By this I mean only to capture some intuitive sense of theoretical levels, rather than invoking any particular account of levels. So, for instance, I take it that intuitively we all agree that microphysics, chemistry, biology, and psychology are all theories at different levels.

It is debatable whether there is any correct translation that maps sentences of thermodynamics onto sentences of statistical mechanics, and similarly in the case of meteorology and physical theory. In the former case it might be objected that in statistical mechanics the second law of thermodynamics is false, while in thermodynamics the second law of thermodynamics is true. In the latter case it might be objected that on some occasions the generalisations of high-level meteorological theory are literally false. In the case of thermodynamics and statistical mechanics one could argue that properly understood in thermodynamic theory the second law is no law at all, but a mere empirical generalisation to the effect that in closed systems entropy will increase from a level of low entropy. That is perfectly consistent with what statistical mechanics tells us (it merely also tells us that in times of high entropy, entropy may decrease). Regardless of whether these particular cases are ones where we have a correct translation, if we suppose that there *are* theories such as these that occur at different levels and which are inter-translatable, it seems plausible that they may not be equally explanatorily powerful.

There are two senses in which theories could fail to be equally explanatory, depending on how one understands the notion of explanatory power. At one end of the continuum we have an understanding of explanatory power that is largely psychological, according to which the explanatory power of a theory is understood in terms of the extent to which it creates understanding in the mind of some relevant group of humans. Thus two theories would be equally explanatory just if they created the same degree of understanding in that same group of persons. This psychological notion of explanation is subjective, and entails that any infinite theory, such as, say, the microphysical theory of meteorology, is less explanatorily powerful than the more usual high-level meteorological theory, since no finite human mind could grasp it. If the notion of explanation is to play any role in helping us to decide whether theories are metaphysically equivalent, then the notion of explanatory power at play is not this highly subjective psychological one. Rather, the understanding we want is a more objective one, perhaps some sort of deductive-nomological account of explanation, or in some cases a causal account.[19]

In the case of, say meteorological and physical theories, if these theories were inter-translatable then it seems that they would still fail to turn out to be explanatorily equivalent. For the infinite physical theory fails to tell us what it is in common between the infinite number of physical states mentioned by any particular disjunction, such that they all count as realising the same meteorological state. Arguably, it is an objective feature of the theories that the meteorological theory has greater explanatory power with respect to this particular aspect. So I think it is at least plausible that where we have theories at different levels, we should not expect those theories to be equally explanatory even if they are correctly inter-translatable and thus metaphysically equivalent.

But what of the more usual cases we have been describing, where we are considering theories that are at the same level? If we rule out a psychological notion of explanation, the question is whether two metaphysically equivalent theories at the same level could differ in explanatory power. A detailed answer to this question would require a considered examination of an objective account of explanation. As I see it though, metaphysically equivalent theories at the same level ought to be explanatorily equivalent in *some* objective sense. No doubt the truth of this claim depends in part on which account of levels one accepts. It requires, for instance, that levels be genuinely objective, that is, that relative to different interests, one and the same theory cannot exist at multiple levels. It does not require, however, that there be a single hierarchy, or a non-branching hierarchy of levels. Nor does it require that theories at higher levels be irreducible to theories at lower levels, or be in some way 'emergent'. Nor does it require that in all cases there is some fact of the matter as to whether two theories are at the same level. Perhaps there is no fact of the matter as to whether a theory of flower pollination and a theory of economics are on the same theoretical level. All that is required is that in some cases we are fairly confident in holding that theories are at the same level. And plausibly, frequently where we are considering putative cases of metaphysical equivalence these are precisely cases where we have such confidence. In part this is presumably because pairs of theories such as nihilism and universalism, three- and four-dimensionalism and so forth, are either equivalent or they are competitor theories aiming to fill a single 'theoretical slot'. If these competitor theories are not equivalent, then at best one can be true. Where such theories are at the same level they are attempting to fill the same explanatory niche. Plausibly then, if they are metaphysically equivalent we should expect them to be equally explanatory.

Then given some general objective account of explanation, we could distinguish a number of senses in which theories at the same level might be equally explanatory.[20] Two theories might be equally explanatory in the sense that the 'amount' of explaining each does is equivalent. Then the idea would be that you tally up the amount of explaining done by one theory, and compare it to the amount of explaining done by the other theory. Alternatively, two theories might be equally explanatory if each explains all of the same data, and neither theory provides a better explanation of any piece of data than the other theory. Or two theories might be equally explanatory if they explain all of the same data, neither theory provides a better explanation of any piece of data than the other theory, and the explanations each theory provides for the data are of the same kind. Now, it seems clear that the first sense in which theories could be equally explanatorily powerful, is not the sense that is pertinent to determining whether they are equivalent or not. For it allows that two theories might be equally explanatory overall, but one theory might explain one set of data better than the other theory, while the second theory might explain a different set of data better than the first theory.

Indeed, I think that even the second sort of explanatory equivalence is insufficient for our purposes. It rules out the case above, for it tells us that if two theories are equally explanatory, then there is no piece of data that one theory explains better than the other. But it would allow that theories could turn out to be metaphysically equivalent, despite furnishing very different kinds of explanation for the same data. Plausibly though, if theories are equivalent, then we should expect it to be the case not only that they are equally explanatory in this second sense, but that the explanations they provide are of the same kind: each theory explains the relevant phenomena in the same sorts of ways. Now, it is hard to get at exactly what it means to talk of the same 'kind' of explanation. The intuitive idea is this. Once we have some sort of mapping between theories, we can then look at the explanatory apparatus of each theory. If the translation is indeed a correct one, then we should see a clear sense in which each theory explains phenomena in the same way. Of course, this loose sense of 'same kind of explanation' might be open to interpretation and debate: there might be disagreement about whether the explanations really are of the same kind. And some of these debates might be hard to resolve without appealing to a rigorous account of explanation. But a good deal of the rest of the work of this book lies in showing that the theories of three- and four-dimensionalism are equally explanatory in this third, strong sense: they provide the same kinds of explanation. The hope is that we can get a strong sense of what it *means* to talk of theories providing the same kind of explanation, by looking in detail at a case where, I argue, despite appearances this is exactly what we find. So the hope is that the rest of the book will demonstrate what it is for theories to provide the same kind of explanation, and hence to be equally explanatory in this sense. Henceforth then, when I talk about theories at the same level being equally explanatory, I mean that they are equally explanatory in the third sense outlined above: they explain the same data, there is no piece of data that one theory explains better than the other, and the explanations each provide for each piece of data are of the same kind.

So what does all this tell us? Well, it suggests that where we have theories at the same level, if we can show that those theories are not equally explanatory then we can conclude that we do not have a correct translation between them. Of course, the reverse is not the case: showing that two same-level theories are equally explanatory will not entail that an assertibility mapping is a correct translation.

Indeed, for fictionalists and Platonists, the debate about equivalence rests almost exclusively on consideration of explanatory power. Thus fictionalists may complain that Platonism is explanatorily lacking, since given that mathematical objects are causally inert, it fails to explain how one could ever have knowledge of mathematical truths, while Platonists may complain that fictionalism is explanatorily lacking since it fails to explain the nature of our ordinary semantics in mathematical discourse.[21] If such arguments are compelling, then we have reason to suppose that if these theories are same-level theories, then they are not metaphysically equivalent.

All this leaves us with a question. If same-level theories that fail to be equally explanatory thereby fail to be correctly inter-translatable, do they also fail to be practically translatable? Is discovering that same-level theories for which there is an assertibility mapping are not equally explanatory, just discovering that a practical translation is not a correct translation, or is it discovering that the assertibility mapping is not truth preserving and thus that we have no practical translation? Either way we are able to rule out the existence of a correct translation. But it is an important question, since if we think that the former is the case, then we have a way of ruling out that some practical translation is a correct translation.

I think that considerations of explanatory power are important both in determining whether we have a practical translation, and if so, whether we have a correct translation. In most cases differences of explanatory power such as those described above will surely signal that an assertibility mapping is not truth preserving. If fictionalism can better explain how we have mathematical knowledge, or if nihilism can better explain our intuitions about persistence and convention, then these substantive explanatory differences suggest that at least some of the assertibility mapping is not truth preserving. It entails, therefore, that we do not have a practical translation. In fact it might be possible to use the details of the difference of explanatory power to locate where the assertibility mapping fails to preserve truth, and then to give some explanation of why the proponent of one theory utters frequent falsehoods, thereby meeting the principle of humanity.

3.4 Diagnosing a Correct Translation: Explanatory Idle Elements

But what of using explanatory considerations in determining whether a practical translation constitutes a correct translation. This brings us back to the issue we considered earlier, where what looks like a practical translation fails to count as a correct translation in virtue of the addition of necessary truths as conjuncts of the sentences of one theory. Of course, in such cases we only have a practical translation if all parties agree that the 'additional' bits of the theory in question—the 'God exists' or the '*p* or not *p*'—are necessary truths: otherwise any assertibility mapping will not be truth preserving.

One way we might try and deal with this problem is via some theoretical constraint of simplicity. We might argue that simplicity is a guide to whether a translation is correct or not, that is, that a practical translation is only a correct translation if the theories are equally simple. What tells us that the sorts of cases we have been discussing are not cases where we have a correct translation, is that one theory is obviously less simple than the other.

The problems with this move are twofold. First, it is not clear why we should think that only equally simple theories are ever metaphysically equivalent. This sounds a little more plausible if we hold that theories at the same level will, if equivalent, be equally simple. Even this claim though, would need to be convincingly argued for before we could use simplicity as a guide to whether a practical translation is correct or not. Second, even if such an argument were forthcoming, there would be momentous practical difficulties in using simplicity in this manner, given that it is notoriously difficult, if not impossible, to provide any formal account of simplicity.[22] Indeed, it may be that there is no objective marker of simplicity, but rather, that claims about the relative degrees of simplicity of theories are all culturally and inter-personally subjective. Those who think this latter will hold that there is no reason to suppose that any of our subjective judgements about the relative degree of simplicity of theories is any guide to their equivalence or lack thereof. Even those who think that there is some objective notion of simplicity, concede that formulating judgements about simplicity is difficult because it is always possible to make a theory *appear* simpler by burying the complexity in the atomic predicates of the language.[23]

I suspect that if there is some objective measure of simplicity, then we should expect that the simplest version of theories at the same theoretical level will be equally simple if those theories are correctly inter-translatable. Even if that is true though, it is arguably the case that determining whether the simplest versions of any two theories are equally simple is more difficult than determining whether or not they are correctly translatable. So whatever one's take on the simplicity issue, it is difficult to see how considerations of simplicity will be of help in determining whether or not theories are correctly translatable.

There is, however, a way in which we can rule out some practical translations as being correct. In the previous section I spoke of 'substantive' explanatory differences between theories. I use this term to distinguish the normal sorts of cases of differential explanatory power, from cases where we conclude that two theories differ insofar as one theory has additional explanatorily redundant elements such as the addition of necessary truths. Since the addition of '*p* or not *p*' to every sentence you assert makes no difference to the truth value of that sentence, and adds nothing explanatorily to the theory in which it is embedded, we can conclude that the theory with these additions is less non-substantively explanatory[24] than its rival. Of course, this is not the sort of substantive explanatory difference we discussed earlier. It is sufficient, however, for us to conclude that where we have a practical translation between theories, if one of those theories is less explanatory in this non-substantive sense, then we fail to have a correct translation. Thus the use of such explanatory considerations rules out as equivalent, theories of this sort which are practically translatable but which are not translatable in virtue of all of the same truth makers. Moreover, where we have theories that are practically translatable but fail to be correctly translatable in virtue of failing to be non-substantively explanatorily equivalent, there is good reason to suppose that there is some 'pared down' version of one of the theories such that that theory and its putative rival are explanatorily equivalent and therefore possibly correctly inter-translatable—just remove all of the '*p* or not *p*' additions and there is a good chance that the remaining theories are equivalent.

4. WHY IS THIS METAPHYSICAL EQUIVALENCE?

So far I have argued that theories are metaphysically equivalent just if they are correctly inter-translatable. Moreover, I have provided a number of diagnostic criteria to aid us in determining whether or not such a correct translation exists. These diagnostic criteria provide necessary but insufficient conditions for the obtaining of the relation of metaphysical equivalence. Thus they provide a way of falsifying some claim about the equivalence of two theories: if it can be shown that some diagnostic criterion is not met, then we know that the assertibility mapping is not a practical translation, or that the practical translation is not a correct translation. Suppose though, we find that in some cases all of the diagnostic criteria are met by two theories. Are we then in a position to claim that we have a correct translation, and thus a case of metaphysical equivalence? Certainly this conclusion is not entailed. We have a correct translation only if both theories have the same truth makers, and meeting the diagnostic criteria does not entail that this is so. So what would licence such a move?

One reason we might conclude that if the diagnostic criteria are fulfilled then we have a case of metaphysical equivalence, is if we are committed to what Hirsch calls a shallow approach to ontology, and what Sidelle refers to as the semantic approach.[25] The idea is that in the sorts of cases we have been considering, once we know all of the facts we see that deciding which of two theories is true is not a deep metaphysical matter, but rather is just a matter of deciding the best way to describe those facts in our language. Or, to put it another way, whether a Doe exists or not is not something that could be made true by some 'metaphysical fact'.[26] Once we know how the various simples are arranged, we know all of the facts, and it is merely a semantic question whether or not, given what we mean in English, a certain description of those facts is a proper one. So in deciding which theory is true, we are merely deciding which fits best with our ordinary talk.

Given this semantic approach, it is easy to see why we would conclude that theories that meet our diagnostic criteria are equivalent: for if theories meet the diagnostic criteria, then they could only fail to be metaphysically equivalent if there were some extra metaphysical facts in virtue of which one theory were true and the other false. That is, they could only fail to be equivalent if there were some unobservable truth makers that are explanatorily redundant. Since the semantic approach rejects the existence of such facts, it follows that proponents of this approach will have every reason to think that theories that meet these diagnostic criteria are metaphysically equivalent.

Reaching this conclusion does not require, however, that one adopt this semantic approach to ontology. It is not necessary that *in general* one deny the existence of such extra metaphysical facts. Rather, the question can become whether on any particular occasion, there is good reason to suppose that there are such extra facts. Consider, there are many unobservable facts. There are unobservable facts about the big bang. There are unobservable facts about the existence or not of other worlds that are causally inaccessible to this world. If modal realism is true, then it is so in virtue of facts that are unobservable to us. Although these facts are unobservable, however, we understand what sort of facts they are. Scientific theories about the big bang, or the inside of a black hole, tell us not only that there are certain facts and that some of these facts are unobservable, but these theories also explain why it is that the facts that are unobservable are so. So too in the case of other possible worlds, although it is impossible for us to observe worlds that are causally isolated from us, we at least have some understanding of what such worlds would be like, and why they are unobservable. In each case proponents of these theories can point out what facts would need to obtain for the theory to be true. Further, broad features of the theory *itself* explain why some of the facts are unobservable. These sorts of unobservable facts are posited by a theory in an integrated, systematic way that provides additional explanatory power to the theory, and where this is an explanation for the unobservable status of the facts.

Suppose though, that we are considering theories that meet the various diagnostic criteria we have outlined. And suppose that someone insists that despite meeting those criteria, these theories are not metaphysically equivalent: there is some further fact that entails that at most only one theory is true. Then we should ask whether this is a fact that is posited by either of the theories in question in an integrated manner. Suppose it turns out that three- and four-dimensionalism are examples of theories that meet these diagnostic criteria. Suppose further that the three-dimensionalist insists that despite this, three-dimensionalism is true, and four-dimensionalism is false. If this is so, then it must be because of the obtaining of some explanatorily idle, unobservable fact. Suppose the three-dimensionalist says the following: this fact is the fact that persisting objects are strictly identical across time. Then it might look as though she is simply appealing to some fact that is indeed part of the explanatory apparatus of her theory. Now, it is certainly true that three-dimensionalists think that persisting objects are strictly identical across time. So one might well think that if there was some genuine difference between a world where objects endure, and a world where they perdure, then this might be in virtue of the fact that in the first world, the objects are strictly identical across time, and in the second world they are not. I agree. If there was a real difference, we might expect it to be located there. Indeed, I take it that three-dimensionalists think that there *are* differences between worlds where objects endure, and worlds where objects perdure. That is, they think that this difference makes a difference: it might not make an empirical difference, but it should make a metaphysical and an explanatory difference. The account we give of an enduring world, should be different to the account we give of a perduring world. So, for instance, three-dimensionalists tend to think that the fact that enduring objects are strictly identical across time, means that the account of property instantiation that they provide must be different to that provided by someone who rejects such a claim. They think that being strictly identical across time has some explanatory and metaphysical consequences. If it turned out that the *only* difference between a world in which objects were strictly identical across time, and a world in which they were not, was that in one world we claimed that there was some additional unobservable and explanatorily redundant fact that simply *made it the case* that strict identity held in that world, then we might be tempted to say that this is no difference at all.

Or alternatively we can imagine some four-dimensionalist making the claim that despite the two theories meeting the diagnostic criteria, nevertheless, four-dimensionalism is true and three-dimensionalism is false. And we can imagine her pointing to the (claimed) existence of temporal parts as being the relevant further fact in virtue of which four-dimensionalism is true. And surely it is indeed part of the apparatus of four-dimensionalism that it posits the existence of temporal parts. Again, this is true. If perdurantism turns out to be true of our world, then this is no doubt because our world is one in which objects are composed of temporal parts. But notice that the four-dimensionalist does not posit the existence of temporal parts *qua* unobservable and explanatorily redundant objects. Four-dimensionalists think that positing temporal parts provides their theory with a good deal of explanatory power, and indeed, an explanatory advantage over three-dimensionalism. They think temporal parts *make a difference*. That's why they posit them! In this case, however, we are considering what we should say if it turns out that three- and four-dimensionalism meet all of the diagnostic criteria. Then the only difference between a world in which one theory is true, and a world in which the other theory is true, is the obtaining of some metaphysical fact that makes no explanatory or empirical difference. We could say that this fact is the existence, in that world, of temporal parts. But it is no part of the theory of four-dimensionalism that temporal parts are explanatorily redundant. It is no part of their theory that the only difference between an enduring world and a perduring world, is that in the latter some fact obtains which makes absolutely no difference except to serve as a truth maker for the claim that four-dimensionalism is true in that world. If, in the end, the only difference between an enduring world and a perduring world is the putative existence of this fact, then surely we have good grounds to think that there is no difference.

This is just to say that in both of these cases we seem to have an *ad hoc* measure of introducing some unobservable fact that provides no additional explanatory power to the theory, purely in order to maintain that the relation of metaphysical equivalence fails to hold. This seems no more than a desperate attempt to hold that there *must be* some fact that determines that one theory is true and the other false.

If we find, on some occasion, that there is a plausible non *ad hoc* explanation for why some relevant truth maker is unobservable and explanatory redundant, then we might feel justified in holding that there is such an extra fact on that occasion. So under certain circumstances we might feel justified in concluding that two theories that meet all of the diagnostic criteria fail to be metaphysically equivalent. In general though, the sorts of cases we have been considering do not appear to be cases where additional metaphysical facts are posited by the competing theories themselves. Rather, the positing of such facts appears to fall into the category of an *ad hoc* measure: these facts add no explanatory power, and are not part of the theory itself. Given this, there seems little reason to suppose that such facts exist. Once we see this though, something like inference to the best explanation should tell us that theories that meet the diagnostic criteria are metaphysically equivalent. What explains the fact that the theories meet all of these criteria? Well, they are actually correctly inter-translatable; they have the same truth makers; they are metaphysically equivalent.

5. HOW DOES THIS ACCOUNT HELP?

It might seem though, that some of these criteria are unnecessary. We could just determine whether there are any observable facts in virtue of which one theory is true and the other false. If we first determine that two theories are empirically equivalent, then there is a further question about unobservable facts: are there any unobservable facts in virtue of which one theory is true and the other false? To determine whether there are unobservable facts we turn to the idea of explanatory *ad hocness*, and ask whether or not the positing of such facts is part of the apparatus of the theory, or merely an *ad hoc* measure. If it is purely *ad hoc*, then we can conclude that the two theories are metaphysically equivalent; if it is not *ad hoc*, then we can conclude that they are not metaphysically equivalent. We then need no recourse to the talk of inter-translatability or relative degree of theoretical virtues.

We might think that such a proposal is a good one. After all, there is considerable controversy about inter-theoretic translation, and more still about the notion of explanatory power. But there are two reasons to reject this proposal. First, consider how we are to ascertain whether the positing of the relevant unobservable facts is principled or *ad hoc*. Suppose I am considering three- and four-dimensionalism. Suppose everyone agrees that the theories are empirically equivalent. Is there some unobservable fact in virtue of which one theory is true and the other false? Well, if four-dimensionalism is true, then it might be because objects are composed of temporal parts. That fact is unobservable, but clearly positing temporal parts it is not *ad hoc*, it is precisely part of the apparatus of the theory. Does this resolve whether or not the theories are metaphysically equivalent or not? No. For we have assumed that the theoretical terms, as they are used by the four-dimensionalist, are the same terms, as they are used by the three-dimensionalist. But if these terms are in part defined by their role in the theoretical apparatus in which they are embedded, then there is no reason to suppose that when the three-dimensionalist denies that there exist temporal parts and thus that objects perdure, that she is denying what the four-dimensionalist is affirming. Perhaps the claim that three- and four-dimensionalism are metaphysically equivalent is the claim that they mean something different by terms such as 'part' such that although they appear to be making contradictory claims, in fact they are not. Whether or not there is some principled unobservable fact in virtue of which four-dimensionalism is true and three-dimensionalism is false, is impossible to determine independent of knowing whether the two theories are inter-translatable.

More important than this though, the various diagnostic criteria are invaluable in determining precisely where parties disagree about whether theories are equivalent or not. At present it is often unclear why it is that proponents of the view that certain theories are equivalent think that they are so, and equally unclear why those who disagree do so. Where does the disagreement lie? We can now see that there are a number of junctures at which parties might disagree. Parties might disagree from the outset about whether or not there is an assertibility mapping between two theories: they might disagree about whether the theories are empirically equivalent or not. Or they might agree that such a mapping exists, but disagree about whether it is truth preserving: one party might contend that we fail to have a practical translation, and that this is so because there are substantive differences of explanatory power. This disagreement might rest on differences the parties have in how they understand explanation. If so, then it is important to be aware of this. On the other hand, parties might agree that two theories are not equally explanatory, but disagree about whether they are at the same level. Thus one party may conclude that the lack of explanatory equivalence does not entail that there is no practical translation since the theories are not at the same level. Then the disagreement is a product of a different account of theoretical levels. Or parties might agree that there exists a practical translation, but disagree about whether there is a non-substantive difference in explanatory power, thus disagreeing about whether the translation could be correct. Finally, parties might agree that all of the diagnostic criteria are met, but one might thump the table and maintain that there is some unobservable, explanatorily redundant truth maker in virtue of which one theory is true and the other false.

Seeing precisely where the nature of the dispute lies, and upon what it is based is important in resolving such disputes, particularly given that they may sometimes rest on additional, sometimes controversial commitments such as to different theories of explanation or to different accounts of theoretical levels. This account of metaphysical equivalence provides the apparatus with which to clarify the nature of disputes, and a framework within which to argue that theories that we might have thought were radically different, are in fact equivalent.

The remainder of this book is concerned with using the framework of this account in order to argue that two apparently rival theories of persistence: three- and four-dimensionalism, are in fact metaphysically equivalent. For the best way to flesh out an account of equivalence is to put it to work: we can best see how the account will go by examining how it deals with real theories. This is a book that considers issues of theoretical diversity by looking at them through the lens of a particular issue in metaphysics. But it is no accident that the metaphysical issue in question—the issue of in what manner objects persist through time—is the one I chose. Not only is this a pressing contemporary issue, but, as I will argue, answering the question of how objects persist through time—and hence whether three- and four-dimensionalism are equivalent—involves resolving a gamut of other key issues in metaphysics: issues regarding the nature of composition, the nature of simples, and ultimately, the nature of objects themselves. So while the following chapters consider a particular claim about theoretical diversity, they do a lot more than that.

In the next chapter I introduce some of the puzzles we discover when we begin to think about objects and their persistence. In particular, I outline four paradigm puzzles. These puzzles provide a flavour for the sorts of issues that are central to the debate about persistence, and they give insight into what is at stake in developing an account of persistence. More than that though, these are puzzles that any plausible account of persistence must resolve. Since the framework I have proposed for determining whether theories are equivalent or not appeals in no small way to explanatory features of the theories in question, these puzzle cases provide an excellent way to examine the explanatory features of each theory. In addition, they allow us to see not only whether each of the theories is able to explain or resolve the puzzling element, but they allow us to compare each theory's account, and determine whether the sort of explanatory apparatus that each brings to bear is of the same kind. Thus consideration of these puzzles allows us to determine whether three- and four-dimensionalism are explanatorily equivalent, and hence whether they meet one of the diagnostic criteria.

NOTES

[1] Ted Sider raises this possibility in his (1999).
[2] Putnam (1987).
[3] Cf. Putnam (1988).
[4] Hirsch (2002).
[5] Sidelle (2002).
[6] Balaguer (1998).
[7] Balaguer (1998) p 179.
[8] Balaguer (1998) p 151.
[9] I will talk of 'theories' being inter-translatable. I understand theories to be sets of sentences (which make various claims about, or purport to describe the world) in some sub-language. Thus the claim that theories are inter-translatable is just the claim that the sentences of the theories are inter-translatable.
[10] Hirsch (2002) p 54.
[11] Colyvan and Zalta (1999).
[12] See the introduction of Sider (2001); Merricks (2001) chapter one, and Van Inwagen (2002).
[13] Quine (1975).
[14] For a discussion of this distinction see Hoefer and Rosenberg (1994).
[15] See Smart (1968); Quine (1960); Hales and Johnson (2003).
[16] See Davidson (1973).
[17] Not everyone agrees. Merricks, (2001; chapter one), holds that the typical non-universalist means by her sentences of the form 'there is an *x*' just what the folk mean by their same claims. Thus for both the folk and non-universalist, 'there exists a statue' does not mean just what the nihilist or near-nihilist means by 'there exist simples arranged statue-wise.' Thus he thinks that the sentences of the folk and the non-universalist are straightforwardly false (although they are 'nearly' true).
[18] See for instance Lewis (1986) pp 212–213; and (1991) pp 8–81, and Sider (2001) pp 220–240, and (2003) and Hirsch (2002) for a response.
[19] These scientific notions of explanation are a little problematic when we are talking of metaphysical theories, since there is no nomological or causal component. Of course, if two theories are equivalent, then they are also equally D-N explanatory, though the reverse need not hold. It seems that to capture some metaphysical sense of explanation that involves more than merely having two theories entail all of the same sentences, we might need to think of metaphysical explanation as some trade-off between power and simplicity or some such. It is not my task here to provide an account of metaphysical explanation however, whatever the best account might be can be plugged into this account of equivalence.
[20] I owe the following suggestion to Sider.
[21] See Benacerraf (1973).
[22] Cf. Sober (1979).
[23] DeVito (1997).

[24] We could deny that there is any notion of non-substantive explanatory *power*, and instead hold that there is some additional constraint at play—a redundancy or idleness constraint. The difference here is merely terminological. We can talk of theories being equally explanatorily powerful but differing with respect to some idleness constraint, or we can talk of them differing in a non-substantive explanatory manner.
[25] Hirsch (2002) p 67 Sidelle (2002) p 137.
[26] I take it that 'metaphysical fact' here is intended to refer to a type of fact that could, (since is it empirically redundant) only be the kind of fact that would obtain if a certain metaphysical theory turned out to be true. That is, metaphysical facts are facts about a world in virtue of how, and only how, that world is metaphysically.

Chapter 2

THE PUZZLES OF PERSISTENCE

When I was thirteen I had long, straight hair and rather prominent and painful braces, which required the removal of four teeth. Years later the braces are long gone, and various other minor anatomical changes have taken place. In addition to such obvious changes, science informs us that each day we lose certain atoms, and gain others. Thus after a period of time a person is composed of completely different atoms than they were previously. Despite all of these changes, there seems little doubt that I am the same person onto whom those callous pieces of metal adhered.

Equally, it seems that there *could* have been changes that I would not have survived. If I had been hit by a bus and left brain dead, then, arguably, I might have ceased to exist. Furthermore, there are cases where it is less clear what I ought to think about whether I wou ld continue to exist or not. If my brain and body had separately been duplicated, and my original brain placed into the duplicate body, and the duplicate brain into my original body, which, if either of the two remaining persons is me? These are questions that pertain specifically to issues of personal identity. Analogous questions, however, can be raised about objects in general. Such questions are ones regarding the *persistence* of objects over time, where an object persists just if it exists at multiple locations in time.

That objects change as they persist presents a number of problems. For an object only changes insofar as there is some single persisting object that has different properties at different times. But if we think that what it takes to be the 'same' object across time is to be strictly identical across time, then it is difficult to see how an object can have different properties at different times and yet be strictly identical at each of those times. Then how can an object both persist and change? This is one major problem at the heart of a cluster of puzzles about persistence and change. In what follows, I outline four paradigm puzzles of persistence, puzzles that we will return to again and again. The four puzzle cases that I describe are: the puzzle of change; the puzzle of fission; the puzzle of temporary coincidence and the puzzle of permanent coincidence.

These puzzles are important. A good deal of the weight of showing that any two theories, particularly metaphysical theories, are metaphysically equivalent, lies in showing that they are equally explanatory. But it is not possible to consider every explanatory feature of each theory of persistence. Plausibly though, since the puzzles as a group raise a number of wide ranging issues, if it should turn out that each theory deals with these puzzles in the same manner, then this is good reason to suppose that not only are the theories explanatorily equivalent with respect to these puzzles, but that they are explanatorily equivalent *simpliciter*.

Thus I begin by describing these puzzle cases. Then in section three I outline what we might think of as the traditional versions of three- and four-dimensionalism, and I consider how each of these theories deals with each of the puzzles. Having explored each of these broad 'solutions' to the puzzles, it will be apparent why three- and four-dimensionalism are seen as rival theories that preserve different intuitions and have different explanatory virtues and vices. It is this appearance that the remainder of the book sets out to dispel.

1. THE PUZZLES OF PERSISTENCE

1.1 Change

A firm intuition we share is that objects change. To say that an object changes is to say that at some time it has some property, and at another time it does not have that property, or vice versa. An object O changes iff there is some property P, such that at some time t O instantiates P, and at another time t^* O does not instantiate P. Yet in order for O to change with respect to P, O must continue to exist *through* these property changes. That is, change requires that there be an object that persists through the changes and is the bearer of these different properties at different times.

But surely what it is for an object that exists at one time to be the 'same' object as an object that exists at another time, is for the object that exists at one time, to be numerically or strictly identical to the object that exists at another time. The identity relation is frequently held to obey two principles formulated by Leibniz:[1] the Identity of Indiscernibles and the Indiscernibility of Identicals. For our purposes, we need only consider the less contentious Indiscernibility of Identicals, which states that:

For all objects x and y, if x is identical to y then for every property F, x has F if and only if y has F.

$$(\forall x)(\forall y): \quad (x = y \rightarrow (\forall \mathrm{F})(\mathrm{Fx} \leftrightarrow \mathrm{Fy})).$$

I will refer to the Indiscernibility of Identicals as Leibniz' Law, though this terminology is sometimes reserved for the conjunction of both of Leibniz' theses.

Now we have a problem. An object persists only if it is strictly identical across time. Yet objects change over time. Suppose there exists a ball that is red at t and blue at t^*. We want to say that the ball persists despite its change in colour, indeed, we want to say that the ball—the very same ball—changes colour. But Leibniz' Law tells us that the thing that exists at t, is identical to the thing that exists at t^*, only if they share all and only the same properties. But the thing that exists at t is red, and the thing that exists at t^* is not red. So there must be two things: a red thing and a not red thing (a blue thing). So the ball at t is not identical to the ball at t^* and we do not have a case of a persisting object with changing properties. Thus identity across time—the foundation of persistence across time—appears to be inconsistent with change across time.

The puzzle is how it can be that a persisting object can be strictly identical across time and yet exemplify different properties at different times. This is the puzzle of change.[2]

1.2 Temporary Coincidence

The second puzzle is what I will call the *problem of temporary coincidence*. Suppose that there exists some lump of clay, call it Clay, which begins its life in lump form, is gradually sculpted into a statue, call it Statue, and is then later rolled flat, at which point the Statue ceases to exist. Statue exists only when Clay exists, although there are times at which Clay exists and Statue does not. So at every time at which *both* Statue and Clay exist, they share all and only the same parts—they are materially coincident at those times. Thus whenever Statue exists, Statue is coincident with Clay, though there are times when Clay exists and is not coincident with Statue: they are *temporarily coincident*. In general, any two persisting objects O and O^* are temporarily coincident just if there is some time t at which O and O^* materially coincide, and some time t^* at which at least one of O or O^* exist, such that O and O* do not materially coincide at t^*.

When Clay and Statue coincide their relation is a close one: they share the same intrinsic properties, and indeed, are indiscernible. We might even be tempted to say that they are identical at those times. Yet that cannot be, for absent some kind of temporally relative identity[3]—identity that holds at certain times and not at others—objects are either identical or they are not. So Statue and Clay must be distinct, for not only does Clay have a longer lifespan than Statue, and thus have properties at those times which Statue does not, but in addition Clay has properties that Statue does not *even at the times when they are coincident.*

Clay can survive events that Statue cannot, and Statue can survive events that Clay cannot. Clay can survive being squashed flat; Statue cannot. Plausibly, Statue can survive gradually having the clay from which it is sculpted, replaced by bronze. But if we suppose that as each tiny bit of clay is removed and replaced by bronze it is thrown away or dissolved in acid, then Clay does not survive the experience. Clay and Statue have different *persistence conditions*—the conditions under which they persist.

The persistence conditions of Clay and Statue seem to in part be the result of how each is related to their constituent matter. Some objects seem more closely tied to their constituent matter than others, and this is reflected in the kinds of changes that each is able to survive. How an object is related to its constituent matter depends on its *essential properties*—the properties of which an object cannot survive the loss. So *P* is an essential property of *O*, if *O* ceases to exist upon failing to instantiate *P*. Thus essential properties are those properties possessed by an object in every possible world in which that object exists. Thus many want to say that Statue is essentially a Statue, and has the persistence conditions of a statue, while Clay is essentially a lump, and has the persistence conditions of a lump.

So even when Clay and Statue are coincident they have distinct properties: Clay has the property of being essentially a lump, and the property of being able to survive being squashed flat, while Statue has the property of being essentially a Statue, and the property of being unable to survive being squashed flat. Hence by Leibniz' Law, Clay and Statue are distinct. The puzzle lies in defining the relation between Statue and Clay.

1.3 Permanent Coincidence

A third puzzle is closely related to the problem of temporary coincidence. It is the *problem of permanent coincidence.* Alan Gibbard introduces this problem in his story of Lumpl and Goliath.[4] One day a sculptor takes two separate lumps of clay and fashions one into the top half of a statue of Goliath, and the other into the bottom half of a statue of Goliath, and when finished, he pushes the two lumps together to simultaneously form a lump—Lumpl—and a statue—Goliath. Some days later the statue/lump is ripped apart into tiny pieces, thus ending the career of both Lumpl and Goliath.

So Lumpl and Goliath exist at all and only the same times, and at those times they share all and only the same matter. Lumpl and Goliath are materially coincident at every time at which they exist: they *permanently coincide*. In general we will say that persisting objects *O* and *O** permanently coincide just if *O* and *O** exist at all and only the same times, and at each of those times they materially coincide.

Here there is an even stronger temptation to say that Lumpl and Goliath are identical. For there is *no* time at which Lumpl exists and Goliath does not or *vice versa*. Lumpl and Goliath always share the same weight, the same shape and the same appearance. How could they fail to be identical? Yet there is good reason to suppose that Lumpl and Goliath are distinct. For there are properties that Lumpl and Goliath do not share: modal properties. Lumpl is a lump of clay with the associated persistence conditions of a lump of clay. Goliath is a statue of Goliath with the associated persistence conditions of a statue. Lumpl has the property of being able to survive being squashed; Goliath lacks this property. So although Lumpl and Goliath *in fact* permanently coincide, they have different persistence conditions and thus different modal properties, which is to say that there are worlds in which Lumpl and Goliath do not permanently coincide. For instance, there is a world in which instead of the statue/lump being torn apart thus simultaneously ending the lives of both Lumpl and Goliath, instead the statue/lump is squashed, ending the life of Goliath but not Lumpl. So too there is a world in which the clay in the statue/lump is gradually replaced by bronze, thus ending the life of Lumpl but not Goliath.

Ultimately, Goliath and Lumpl have distinct modal properties, and so by Leibniz' Law Lumpl and Goliath must be distinct. The puzzle is how to reconcile the fact that Lumpl and Goliath have different modal properties and must therefore be distinct, with the fact that they are materially coincident at all times at which they exist and therefore share all of the same intrinsic properties, thereby looking for all the world as though they are identical.

Finally then, I turn to consider our fourth puzzle case: fission.

1.4 Fission

Fission occurs when a single individual 'splits' into two qualitatively identical individuals both of whom are related in the same manner to the original individual. Thus, for instance, amoebic asexual reproduction is an example of fission in the natural world. Or consider the case of Star Trek's Will Riker who undergoes teletransportation, which is supposed to instantaneously transport him from one location to another by destroying his body at one end and creating a physical and psychological duplicate at the other end. (Let us also suppose that the physical continuity theory of personal identity is false, and thus that Riker does indeed survive transportation procedures such as this one. Those who cannot abide such radical physical discontinuity can feel free to supplement their own example here.) Suppose that there is a transporter accident resulting in two persons arriving at the destination end of transportation, each qualitatively identical to Will Riker prior to transportation. Call the two post-fission Rikers R_1 and R_2. Then there are three possibilities. Either both R_1 and R_2 are Riker, neither R_1 nor R_2 are Riker, or one and only one of them is Riker.

Suppose we think that neither R_1 nor R_2 is Riker. This seems peculiar. If the normal transporter process had occurred and only one Riker had appeared at his destination, then Riker would have survived the procedure. But how could the fact that two qualitatively identical—and hence psychologically continuous—Rikers appeared, mean that neither of the resulting persons is Riker? In that case identity is extrinsic, for whether or not Riker continues to exist depends on extrinsic facts about whether or not a transporter incident creates two psychologically continuous persons rather than one. Yet surely identity is not extrinsic, and whether Riker survives or not cannot depend on how many psychologically continuous persons are created at the destination end of transportation.

Yet if we think that Riker did survive transportation, then it seems that only one of R_1 and R_2 can be Riker. Identity is transitive: if A is identical to B, and B is identical to C, then A is identical to C. So if Riker is identical to R_1, and Riker is identical to R_2, then by transitivity R_1 is identical to R_2. But R_1 and R_2 are not *strictly* identical, they are merely *qualitatively* identical. Thus Riker cannot be identical to both R_1 and R_2.

Thus we are left with the option of saying that either R_1 but not R_2 is identical to Riker, or that R_2 but not R_1 is identical to Riker. But on what grounds could R_1 be identical to Riker and R_2 fail to be identical to Riker, or *vice versa*, given that R_1 and R_2 are qualitatively identical and each is related to the original Riker in exactly the same manner? Whatever relation holds between R_1 and Riker will also hold between R_2 and Riker. So there can be no principled reason to hold that one, rather than the other of the resultant persons is identical to Riker.

So we have competing intuitions. On the one hand it seems that Riker survived the transporter procedure. On the other hand it seems that there is no principled reason why one of the resulting persons, R_1 or R_2, is identical with Riker and the other is not. Again then, we find that persistence and change throw up challenges in the face of identity.

These then, are the four paradigm puzzles. Now that I have described each of them, I turn, in the next section, to outline two general approaches to persistence—three- and four-dimensionalism—and then I proceed to examine how each of these theories, broadly construed, solves, or attempts to solve, these four puzzles.

2. TWO GENERAL APPROACHES TO PERSISTENCE

2.1 Three-dimensionalism

Broadly understood, three-dimensionalism is the thesis that objects persist through time by being wholly present at every temporal location at which they exist, and identical to themselves at each of those times.[5] Three-dimensionalists hold that objects are extended across only spatial dimensions, and thus if space has three dimensions, then (composite) persisting objects are three-dimensional. Then three-dimensional objects persist through time by *enduring*. The basic idea is clear enough. The three-dimensionalist is committed to a conception of persistence according to which objects wholly exist at progressive temporal locations, and as such can be seen as 'moving through' time. This is in contrast to the broadly four-dimensionalist conception, according to which objects are extended not only across the spatial dimensions, but also across a temporal dimension.[6] Thus if there are three spatial dimensions, they hold that persisting objects are four-dimensional. Four-dimensional objects are extended in time as well as space—they are literally 'spread out' in time, and are never wholly present at any temporal location. Rather, they exist at different times by having (temporal) parts at those times.

In chapters three and four I elucidate more thoroughly the theories of three- and four-dimensionalism, disentangling a number of related issues and arguing that there are a number of quite different versions of each theory, almost none of which look like 'traditional' three-dimensionalism, and some of which look nothing like 'traditional' four-dimensionalism—the two views that I am about to describe. For now, however, I want to consider traditional three- and four-dimensionalism and examine the solutions that these views offer to the puzzle cases.

2.2 Broadly Three-dimensionalist Approaches

In what follows I consider the manner in which a traditional three-dimensionalism, broad construed, deals with the various puzzle cases. I begin with the puzzle of change.

2.2.1 Change

As we noted in the previous section, one of the major puzzles of persistence is the problem of change, which arises out of an apparent inconsistency between a persisting object changing over time and being strictly identical across time. Suppose that a ball changes colour by being completely red at t_1 and completely blue at t_2. It seems that the ball at t_1 cannot be strictly identical to the ball at t_2, since at t_1 the ball has the property of being red, and at t_2 it lacks this property. To put it another way, the ball cannot straightforwardly have the property of being wholly red and the property of being wholly blue, for these are contradictory properties. Yet if the ball is numerically identical across time it has just one set of properties, and these properties must be consistent.

Three-dimensionalists solve this problem by either relativising properties to times—a view known as indexicalism[7]—or temporally relativising the manner in which properties are instantiated—a view known as adverbialism.[8] This is where the problem of change often turns into the problem of temporary intrinsics. For the claim is that at least with respect to indexicalism, this means that we must treat all properties as relational: they are disguised relations to time. Thus there is a problem of temporary *intrinsics*, because it looks as though the indexicalist cannot make sense of the idea of a truly temporary intrinsic. We will consider this issue later. For now we need only see how these strategies are supposed to resolve the problem of change.

Then, according to the indexicalist, the ball has the properties of being red-at-t_1 and blue-at-t_2, The ball's instantiating these temporally relativised properties is consistent with the ball being strictly identical across time, for it can instantiate without contradiction the properties of being red-at-t_1 *and* blue-at-t_2 at every time at which it exists, and this is consistent with the ball's being strictly identical across time, and also with it changing across time.

In contrast, the adverbialist solution modifies the *manner* of instantiating a property rather than the property itself. So properties themselves are not disguised relations to times, rather, the instantiation of a property involves a relation to a time: the instantiation relation is temporally relativised. Thus given adverbialism, a property that is instantiated at t, is instantiated in a $t$ly manner. So the ball will tenselessly have the properties of being red $t_1$ly and blue $t_2$ly. Once again, there is no contradiction in the ball having the properties of being *both* red $t_1$ly and blue $t_2$ly. Hence the ball's having all of the same 'adverbialised' properties at every time at which it exists, is perfectly consistent with the ball being strictly identical across time. Change and strict identity are reconciled.

Let us now proceed to consider the puzzles of both temporary and permanent coincidence, before turning, in the next section, to consider the puzzle of fission.

2.2.2 Temporary and Permanent Coincidence

Three-dimensionalists generally treat cases of temporary and permanent coincidence in an analogous manner. They maintain that there exists a relation that holds, at a time, between distinct objects that are materially coincident at that time. So, for instance, there is some relation that holds between Statue and Clay, and between Lumpl and Goliath at the times at which these objects materially coincide. This is the same relation that holds between persons and their bodies, between legal tender and pieces of paper and so on. It is a relation to be found in abundance. It is the relation three-dimensionalists call *constitution*.[9] Constitution may sometimes appear like identity because objects related by constitution at a time, share all of the same intrinsic properties at that time. But such objects are not identical because they do not share the same essential properties, persistence conditions or modal properties.

The introduction of the constitution relation allows the three-dimensionalist to resolve the puzzles of both temporary and permanent coincidence. In the case of temporary coincidence, she can maintain that although Statue and Clay are not identical—for they instantiate different properties—they are closely related at times, namely all of the times at which Statue exists: for at those times Statue and Clay are related by constitution.

Similarly, Lumpl and Goliath are also related by constitution, but they are related by constitution at *every* time at which they exist—that is, there is no time at which Lumpl exists and is not related by constitution to Goliath at that time and *vice versa*. So in cases of permanent coincidence there is an even greater temptation to think that the objects in question are identical—for such objects are always related by constitution and thus always share the same intrinsic properties. Yet they are distinct, and invoking the constitution relation allows us to explain how such objects can have different modal properties—for it is possible that Lumpl and Goliath might have been related by constitution at only some of the times at which they exist (there are worlds in which they are temporarily coincident) and it is possible that Lumpl might have constituted some statue other than Goliath (there are worlds in which Lumpl is related by constitution to a statue other than Goliath).

But what does it mean to say that two objects are related by constitution, or that one object constitutes another? There are a number of different three-dimensionalist accounts of the constitution relation, though almost[10] all agree on the basic claim that it is the relation that holds between distinct enduring objects at the times when those objects materially coincide. It will not be possible to give a detailed consideration of any of these accounts here. What almost all of the accounts share though, in addition to the basic claim, is the desire to capture some transitive, irreflexive and asymmetric relation that tracks our intuitions about when to say that one object constitutes another. So, for instance, many three-dimensionalists want to say that the lump of clay constitutes the statue, that the paper constitutes the dollar note, that the body constitutes the person, but that the reverse is not the case.[11] So for many three-dimensionalists, the constitution relation is not *just* the basic relation that holds between any two materially coincident objects at a time, but rather, it relates these objects in a particular way. For instance, Baker[12] holds that the way in which objects are related by constitution creates an ontological hierarchy such that the object that is constituted is more ontologically significant than the object that does the constituting. Since an object can be both a constituter and a constituted, we have a hierarchy of constitution relations.

On most views, what determines which of two objects is the constituter and which the constituted depends in some way on the modal properties of the objects in question. For Baker, broadly speaking an object O constitutes another object O^* at t iff at t O and O^* spatially coincide, and although it is possible that the constituting object O existed at that time without constituting O^*, as it happens at t O is in O^*-favourable circumstances and thus necessarily constitutes O^* at t. Thus if there exists a lump of clay that is in statue-favourable circumstances, then necessarily there will exist a statue that is constituted by the clay, though the reverse is not true since the statue might have existed at that time but not have been made from clay. So we say that the clay constitutes the statue rather than the reverse. For Thomson, the asymmetry of constitution is the result of constituting objects being more tightly tied to their parts than constituted objects.[13] The idea is that the clay constitutes the statue because there is some part of the clay that is an essential part of the clay but which is not an essential part of the statue (say the finger) but there is no part of the statue that is an essential part of it, but which is not also an essential part of the clay. Along similar lines, Doepke[14] and Simons[15] have an analysis of constitution which focuses on the idea that the constituted object is dependent on the constituting object for its existence. Their idea is that while the constituting object could serve as a substratum for the constituted object's destruction, the reverse is not the case.

All of these accounts face various purported counterexamples where folk intuition would tend to say that A constitutes B rather than the reverse, but on the analysis provided we must conclude that B constitutes A. As I see it, these accounts fail largely because they are attempting to integrate into an analysis of the constitution relation, various folk intuitions about how we use the word 'constitutes' in English, and it is unclear that such intuitions are sufficiently systematic or coherent for it to be either possible or desirable that an analysis so integrate them.

In chapters four and five I will discuss and provide an analysis of the constitution relation within the context of various views about composition. For now though, by 'the constitution relation' I mean just the basic relation that holds between any two materially coincident three-dimensional objects at the time of material coincidence. As is stands, this relation is not asymmetric—if *A* constitutes *B* at *t*, then *B* also constitutes *A* at *t*, or, to put it another way, both *A* and *B* are related by constitution at *t*. Understood in this manner, though the constitution relation might not capture all that we intuitively mean by our English term 'constitution', it is sufficient to resolve the relevant puzzles. Cases of temporary and permanent coincidence are ones in which the objects in question are related by constitution at some or all of the times at which they exist, thus explaining both why those objects can appear, at times, to be identical, and also why they are in fact distinct.

2.2.3 Fission

Finally we turn to consider the puzzle of fission. As we noted previously, the difficulty posed by cases of fission is that it seems impossible to reconcile the intuition that the pre-fission object survives, with the intuition that there is no principled reason why one but not the other post-fission object is identical with the pre-fission object. Some three-dimensionalists simply bite the bullet here, and hold that, *pace* our intuition to the contrary, only one of the post-fission objects is identical to the pre-fission object, but we simply do not know which. Wiggins, for instance, holds that when amoeba undergo fission, something about their nature will determine which of the two resulting post-fission amoebas is identical with the pre-fission amoeba, it is just that this information is epistemically inaccessible to us.[16]

That this proposal rejects one of our intuitions is not in and of itself a reason to reject the proposal—for perhaps the only way to analyse fission will mean the rejection of one or more inconsistent intuitions—however, as Robinson has argued, it does seem implausible to think that given what we know about amoebic fission, anything in the amoeba determines which of the post-fission amoebas is identical to the pre-fission amoeba.[17] That is, it is not just that we do not know what the relevant fact at work here is, but rather, it is difficult to imagine just what that fact could possibly be. It is easy to imagine that we might know all there is to know about the internal workings and reproductive cycle of amoebas, and still have no idea which of the post-fission amoebas is identical to the pre-fission amoeba.

Another proposal involves adopting a simple version of the best candidate theory,[18] according to which where we have two competitor candidates vying to be identified with a particular object, we should choose the best candidate. Of course, the problem here is that there is no single best candidate. We cannot say, by transitivity of identity, that both candidates are identical to the pre-fission object. So it seems that a best candidate theory must say that in these cases, neither of the post-fission objects is identical to the pre-fission object. What we can say in cases where both candidates have equal claim to identity, is that either of them *would* have been identical to the object in question, if the other had not existed. This would allow us to preserve the intuition that had only one Riker emerged from the transporter, then Riker would have survived the transportation procedure. And that seems certain enough. We will say that if R_1 had existed but not R_2, then R_1 would have been identical to Riker, and if R_2 had existed but not R_1, then R_2 would have been identical to Riker. However, although this view allows us to say that had things gone differently Riker would have survived transportation as, say, R_1, this view too requires that we reject one of our central intuitions, since it requires us to say that Riker did not *in fact* survive the procedure. So the best candidate theory not only involves the rejection of the central intuition that Riker did survive transportation, but it also means that, somewhat implausibly, identity is an extrinsic relation: an identity relation between Riker and R_1 would have existed had R_2 not also existed, even though the intrinsic relation between Riker and R_1 would have been the same as it in fact is.[19]

Another way around the fission problem would be to follow a broadly Parfitian line and argue that what fission shows us is that it is not the identity relation that matters when it comes to survival.[20] So we might accept that given the transitivity of identity, it follows that neither R_1 nor R_2 are identical with Riker. But, we might argue, whether identity holds is not relevant to determining whether Riker survived the transportation procedure or not, since it is not the identity relation that determines whether some object 'survives' through some period of time. Of course, 'survives' in this sense is a technical term: I survive just in case there is some future person towards whom I have certain attitudes of care: a person whose existence matters to me, whose pains I want to avoid, and so forth. Then I might survive in this sense, just if there is some future person who is psychologically continuous with me, even though that person is not identical with me. So for Parfit, since both post-fission persons are psychologically continuous with pre-fission Riker, and since it would be rational for pre-fission Riker to be concerned for the welfare of *both* post-fission persons, this is sufficient to show that psychological continuity is what matters in 'survival,' not whether the identity relation holds or not.

Parfit's account has the virtue of upholding both sets of intuitions. For we get to say that Riker does survive, in that he has a psychological continuer. We also get to say that since what is relevant to survival is psychological continuity, and both of the post-fissions persons are equally psychologically continuous with pre-fission Riker, then both have equal claim to be continuers of Riker. So Riker survives as both R_1 and R_2.

This view has the disadvantage that it relinquishes identity as determining whether an object survives or not, and this does seem to be a little problematic. We would naturally tend to think that an object survives just if it is one and the same object, that is, just if it is numerically identical to itself at some earlier time. Survival without identity is at least a little perplexing.

We have one final proposal. Denis Robinson[21] has argued that the three-dimensionalist can make use of a strategy analogous to the one employed by the four-dimensionalist, a four-dimensionalist strategy that we will discuss shortly. Robinson points out that the three-dimensionalist can hold that prior to fission, there exist two materially coincident persons, R_1 and R_2, who are related by constitution at all times prior to fission. Hence prior to fission, the name 'Riker' is ambiguous between referring to R_1 and R_2. The reason we imagine, prior to fission, that only one person exists, is that when objects of the same kind (two persons in this case) are related by constitution, we are even more likely to be fooled into thinking there is just one object present. It is not until after fission occurs that we come to realise that there were actually, in some sense, two persons present prior to fission: R_1 and R_2.

But in what sense were there two persons present prior to fission? After all, there is surely something monstrously counterintuitive in the claim that two persons existed prior to fission. Robinson attempts to ameliorate this counterintuitive consequence by arguing that we do not always count 'by identity', rather, sometimes we count 'by constitution'. The idea is that we are counting by identity if we count as one at t, objects that are numerically identical. In contrast we count by constitution at t if we count as one at t, objects that are related by constitution at t. So if prior to fission someone points to Riker and claims that there exists only one person, there is a sense in which this is indeed the case, namely the sense in which it is true if we are counting persons by constitution and not identity. And given that the constitution relation is a close one, insofar as objects that are related by constitution at a time share all of the same intrinsic properties at that time, that we should count by constitution prior to fission seems sensible. After all, when we are counting persons at a time we are generally trying to count the number of beings that have distinct consciousnesses and that are the appropriate bearers of distinct rights and responsibilities, and there is only one of these present prior to fission.

Given the Robinson analysis, we can think of R_1 and R_2 as two persons who are related by constitution at all times prior to fission, and where the fission event 'pulls them apart', so to speak, so that after fission R_1 and R_2 are no longer related by constitution. The sense in which Riker survives, is the sense in which both of the objects to which 'Riker' ambiguously refers prior to fission, survive. R_1 prior to fission is identical to R_1 after fission, and R_2 after fission is identical to R_2 prior to fission. Thus R_1 and R_2 both survive, and it is the identity relation that determines that survival. Hence this account allows us to preserve both sets of intuitions: Riker does survive, and there is indeed no principled reason why one but not the other post-fission person is identical with pre-fission Riker. Not only that, but the account preserves these intuitions whilst also holding that it is identity that matters in survival.

It all sounds pretty heartening. I think, however, that this proposal is faced with problems. Consider: since three-dimensional objects are wholly present whenever they exist, there must be some fact of the matter prior to fission, regarding whether there exist two persons who are related by constitution at that time. How many persons coincide at a time given that at some future time, a fission event will occur? Two (if we have a case of 'single' fission as in the Riker case above). How many persons coincide at a time given that at some future time, no fission event will occur? We might be tempted to say 'one'. Or at least, intuitively we no doubt think that the answer ought to be one. Now, if we think that past, present and future facts are all equally fixed (there is no open future), then we might say that there is only one person present given that no future fission occurs, and this is determined by the totality of facts, including the future facts about the absence of fission. So the issue of how many coincident persons now exist is only *epistemically* inaccessible to us, since we do not know whether in the future some fission event will occur, and indeed, how many post-fission persons will exist given that fission does occur—after all, fission need not only result in two qualitatively identical persons, there could be any number of such persons. If in the future I undergo a fission event and seven qualitatively identical post-fission persons come into existence, then it turns out that prior to fission, seven coincident persons existed. If no fission event occurs in the future, then there exist no coincident persons.

While I grant that the three-dimensionalist might say something like this, notice that it looks a lot like an appeal to backwards causation. Whether or not some future fission event occurs determines how many *wholly present* coincident persons exist in the present. Now, I do think that there are ways that the three-dimensionalist might respond to this argument, and we will consider those in the following chapters. For now we should only note that there is at least something *prima facie* odd about the idea that future events determine how many objects wholly exist in the present. But if future events (fission or the absence thereof) do not determine how many coincident objects exist in the present, then given that fission in the future is *possible*, it must be the case that those coincident objects already exist in the present—otherwise, if fission does occur, those coincident objects will not exist to provide the basis for the analysis. But this looks very bad indeed. Since for any object it is possible that it undergoes a fission event in the future, it will always be the case that there exist multiple coincident objects. Moreover, since the fission event might produce an infinite or very large finite number of post-fission objects, then it must be the case that an infinite or large finite number of coincident objects exist. And these objects exist regardless of whether a future fission event occurs or not: if no future fission occurs, those objects simply permanently coincide.

Suppose then, that things had gone a little differently and Riker had not undergone fission. In that case R_1 and R_2 would still have existed, but R_1 and R_2 would have been related by constitution at *all* times at which they exist (though of course we are oblivious to this in the general run of things). So just as Lumpl and Goliath are actually distinct but materially coincide at all times, so too the same would have been true of R_1 and R_2. If the actual world had been like this, then it would have been true of R_1 and R_2 (in the same way that it is true of Lumpl and Goliath), that there are counterfactual worlds in which R_1 and R_2 only temporarily coincide, that is, worlds in which some fission event does occur. Thus for every name '*N*' that refers to some object that undergoes fission in some possible world *w*, there must exist in every world in which '*N*' refers, multiple coincident objects $O_1 \ldots O_n$ such that '*N*' refers ambiguously to $O_1 \ldots O_n$. And in many of these worlds $O_1 \ldots O_n$ permanently coincide. Plausibly though, we do not want it to be the case that there exist multiple permanently coincident objects. And if that is so, then as it stands the Robinson analysis of fission cannot be the correct one.

So it looks as though although the three-dimensionalist has a number of different strategies for dealing with cases of fission, none of them successfully allow us to reconcile both sets of intuitions regarding the survival of the pre-fission object, with the notion that it is identity that matters in survival.

So far then, we have seen how three-dimensionalism solves, or attempts to solve, each of the four puzzle cases. In the following section I turn to consider how four-dimensionalism fares at the same task. As we shall see, it appears, at least *prima facie*, that four-dimensionalism offers quite different solutions to the puzzle cases.

2.3 Four-dimensionalism

Four-dimensionalism, broadly construed, is the thesis that objects are both spatially and temporally extended. By far the most common version of four-dimensionalism holds that objects are temporally extended in the same manner that they are spatially extended: by having parts at times the way they have parts at locations. This is the view that objects persist by perduring, that is, by having distinct temporal parts at temporal locations. Following standard terminology, I refer to this view as perdurantism.[22] Perdurantists then, construe the problem of persistence over time in terms of the need to specify for each object or kind of object, how each of the distinct temporal parts of that object are united, that is, what particular relation it is (usually a causal or similarity relation) that holds between the temporal parts.[23]

So for the perdurantist, no persisting object is wholly present at any moment in time, any more than a spatially extended object is wholly present at any point in space. Rather, four-dimensional objects exist at each temporal location in virtue of the existence of some instantaneous temporal part at that location. These instantaneous temporal parts are momentary and do not themselves persist, instead, four-dimensional objects persist in virtue of being composed of a succession of these instantaneous temporal parts.

Perduring objects, then, are the mereological sums or fusions of temporal parts. This is sometimes expressed as the idea that persisting objects are 'space-time worms.' Just as a biological worm has a series of spatial segments that are related to each other spatially (in a worm-like manner), so too perduring objects have temporal segments that are related to each other temporally. And just as one end of a biological worm is located in one spatial location in virtue of the existence of some spatial segment at that location, and the other end of the worm exists in a different spatial location in virtue of a different spatial segment existing at that location, so too one end of a four-dimensional space-time worm is located in one temporal location in virtue of the existence of some temporal segment at that location, and the other end of the space-time worm is located in a different temporal location in virtue of a different temporal segment existing at that location. For now then, let us say that an instantaneous temporal part of a perduring object is the intersection of that four-dimensional object and a time, while an extended temporal part is the intersection of a four-dimensional object and a temporal interval. Thus at every time at which they exist, temporal parts share, at that time, all and only the spatial parts at that time of the four-dimensional whole of which they are temporal parts.

I will further define the notion of a temporal part in chapter three, and then go on to consider a number of different versions of four-dimensionalism in chapters five and six. For now, however, this broad explication of four-dimensionalism in its traditional form of perdurantism should be sufficient for an understanding of how, in general, traditional four-dimensionalists deal with each of the puzzle cases.

2.4 Broadly Four-dimensionalist Approaches

2.4.1 Change

Recall that there is a puzzle about how persisting objects change over time. Leibniz' Law tells us that an object considered at one time is identical to an object considered at another time, only if that object has all of the same properties at each of those times, and that seems to rule out the possibility of change over time.

Perdurantists deal with the problem of change by noting that temporary properties are properties of temporal parts. Just as (most)[24] spatially extended objects exemplify local intrinsics at spatial locations in virtue of having spatial parts with those properties at those locations, so too temporally extended objects exemplify temporary properties at times in virtue of having temporal parts that exemplify those properties at those times. Thus objects change by being composed of a series of temporal parts each of which has different properties. Just as spatially extended objects are self-identical, so are four-dimensional objects. Yet just as no two spatial parts of a spatially extended object are identical, so too no two temporal parts of a four-dimensional object are identical.

In the case of the ball we introduced earlier, the four-dimensionalist can admit that the object that wholly exists at t_1 and is red, and the object that wholly exists at t_2 and is blue, are indeed distinct objects just as Leibniz' Law tells us. Each of these objects is a distinct instantaneous temporal part of the ball—a distinct ball-part. The ball changes over time in virtue of having these distinct temporal parts at different times. Thus the four-dimensionalist reconciles both Leibniz' Law and change over time. Moreover, the four-dimensionalist argues, she has no problem with temporary intrinsics in the way the three-dimensionalist does, since she does not appeal to any apparatus that relativises properties to times, and hence there is no danger that all properties will turn out to be relational.

2.4.2 Temporary and Permanent Coincidence

So how do four-dimensionalists make sense of temporary and permanent coincidence? For the four-dimensionalist, both temporary and permanent coincidence are analysed in terms of overlap, where *A* overlaps *B* just if there is some *C* such that *C* is part of *A* and part of *B*. Consider first temporary coincidence. Just as we understand (partial) spatial overlap in terms of the existence of two spatially extended objects that share some proper spatial part, so too the four-dimensionalist understands temporary overlap in terms of the existence of two four-dimensional objects that share some proper temporal part—that is, a temporal part whose temporal extent is less than the extent of the four-dimensional whole of which it is a part.

If we conceptualise persisting objects as space-time worms, then temporary overlap occurs when two worms overlap: when they share the same temporal parts for a period of time. Consider again the case of Statue and Clay. The four-dimensionalist will describe this as a case in which we have some four-dimensional object—Clay—that is the mereological sum of temporal parts that are arranged lump-wise, that is, where the causal or similarity relation that holds between the temporal parts is one that preserves lumpness. There is also another four-dimensional object—Statue—that is the mereological sum of temporal parts that are arranged statue-wise—the relation that holds between the temporal parts and preserves statueness. Some of the temporal parts of Clay are also temporal parts of Statue, and all of the temporal parts of Statue are temporal parts of Clay.

So what it is for Clay and Statue to coincide at *t*, is for them to share a temporal part at that time. Clay and Statue appear to be identical at *t* because at *t* the part of Clay that exists—the t-part—is identical to the part of Statue that exists at t—the t-part. So there really is only one object that wholly exists at *t*—the instantaneous t-part: Clay-at-*t* is identical to Statue-at-*t*. But there also exist (at least) two persisting objects that do not *wholly* exist at that time—Clay and Statue—each of which is differently related to that instantaneous temporal part at *t*. Both Clay and Statue exist at *t* in virtue of the existence of that instantaneous temporal part at that time, but Clay is lump-related to that and the members of set *S* of temporal parts, while Statue is statue-related to that and the members of set *S** of temporal parts. It is these different relations that explain the different persistence conditions of Clay and Statue. For *S* and *S** are distinct: some of the lump-related temporal parts are not also statue-related temporal parts, thus there are times when Clay and Statue do not share temporal parts, and hence times when they do not coincide. Ultimately then, the four-dimensionalist holds that

Statue and Clay overlap, and at the times at which they overlap they share the same intrinsic properties, namely the intrinsic properties of the shared temporal parts that exist at that time.

The case of Lumpl and Goliath is a little more complex. Lumpl and Goliath permanently coincide, thus, for the four-dimensionalist, they *completely overlap*. Let us say that *A* and *B* completely overlap just if there is no particular *P* that is part of *A* and not part of *B*, and no particular *P** that is part of *B* and not part of *A*. Then, for the four-dimensionalist, Lumpl and Goliath share all of the same temporal parts. But just as mereology tells us that if *x* and *y* are spatially extended objects that share all and only the same spatial parts, then *x* and *y* are identical, so too it tells the four-dimensionalist that if *x* and *y* are temporally extended objects that share all and only the same temporal parts, then *x* is identical to *y*. If that is so, then we should conclude that Goliath is identical to Lumpl. Some four-dimensionalist draw this conclusion.[25] They argue that the sorts of modal properties that appear to distinguish Lumpl and Goliath are in fact not genuine properties of objects at all, and therefore there is no reason to suppose that Lumpl and Goliath are distinct. There is no permanent overlap, there is just identity.

Those who hold that the relevant modal properties *are* genuine properties of objects, however, are pushed towards some form of contingent identity. We will discuss the specifics of contingent identity further in chapter five. For now, we should note that contingent identity is, in broad strokes, the thesis that a single individual may have two different designations in the actual world which, under different circumstances, might have referred to two distinct individuals. The contingent identity thesis, then, is the thesis that *statements* of identity can be contingent. Identity itself is necessary. It will always be necessary that *x* is identical to *x*, that is, that *x* is self-identical. What is contingent is that the designation '*y*' picks out *x* in the actual world. So what might have been two distinct individuals is, in the actual world, one individual whose dual designations are contingently identical.[26]

In the case of Lumpl and Goliath, the four-dimensionalist may say that Lumpl is contingently identical to Goliath: the names 'Lumpl' and 'Goliath' pick out the same individual in the actual world, but there are counterfactual worlds in which 'Goliath' picks out an object and 'Lumpl' picks out an object, and those objects are distinct. So recourse to contingent identity dissolves the Lumpl and Goliath puzzle by acknowledging both sets of intuitions: it is the case that Lumpl and Goliath are identical in the actual world in virtue of having all and only the same temporal parts, but it is also true that they are distinct in counterfactual worlds, worlds in which they have different temporal parts.

2.4.3 *Fission*

Finally then, let us turn to consider fission. I noted earlier that Robinson proposes a three-dimensionalist solution to the problem of fission that is analogous to the traditional four-dimensionalist solution. Though I argued that Robinson's proposal faces some objections, the four-dimensionalist analogue of this proposal does not suffer the same objections.

The idea is that four-dimensionalists understand fission in terms of the 'forking' of a space-time worm. Post-fission there exist two distinct objects with distinct temporal parts. Prior to fission those two post-fission objects share the same temporal parts. Fission, then, is a particular instance of temporary overlap, but it is an instance in which each of the overlapping objects is related in the same manner to each of its temporal parts—in the case of Riker, being person-related—such that when the objects cease to share temporal parts, each is related in the same manner to the pre-fission shared temporal part.

Recall that in the case of Riker, after transportation Riker splits into two post-fission persons, R_1 and R_2. Let us call the extended temporal part of Riker that exists prior to fission, *the Riker-part*. Then let us call the extended temporal part of one of the post-fission Rikers, *stage*$_1$, and the extended temporal part of the other post-fission Riker, *stage*$_2$. Stage$_1$ and stage$_2$ come into existence after Riker's transportation accident, at the same time that Riker-part ceases to exist. The four-dimensionalist will describe the case in the following way: there exists some temporal part, the Riker-part, that is a temporal part of two space-time worms. One space-time worm includes (is the fusion of) the Riker-part and stage$_1$, while the other space-time worm includes the Riker-part and stage$_2$. The first space-time worm is R_1, and the second is R_2.

When we point to the temporal part that exists prior to fission—the Riker-part—and say that it is Riker, we are not pointing to a unique individual, we are pointing to a part of both R_1 and R_2. So prior to fission, the name 'Riker' refers ambiguously to R_1 and R_2. The sense in which Riker survives fission is the sense in which both R_1 and R_2 survive: both exist prior to fission and both exist post-fission. This too, is the sense in which there really are no grounds for thinking that one, but not the other post-fission Riker, is identical to pre-fission Riker. Moreover, this analysis also allows us to hold that it is identity that matters in survival. For although neither stage$_1$ nor stage$_2$ is identical to Riker-part, (thus not threatening the transitivity of identity), nevertheless R_1 is strictly identical to itself, and R_2 is strictly identical to itself. So it is identity that matters in survival, insofar as R_1 survives by being identical to itself, and so too *mutatis mutandis* for R_2.[27]

This four-dimensionalist analysis of fission does not share the problem of its three-dimensionalist counterpart. Four-dimensionalists hold that persisting objects are temporally extended, and only partly present whenever they exist. Thus it follows that whether some current temporal part is a part of one or more persisting objects will depend on both past and future facts. Just as we can only determine whether some spatial part is a part of multiple spatially extended objects when we can see all of the relevant objects—for instance, we can only tell whether some wall is part of one house or two when we can see both relevant houses—so too we can only determine whether some temporal part is part of multiple temporally extended objects when we can see all of the four-dimensional objects in question. Thus whether at some time there exists one person, or two materially coincident persons, naturally depends on facts about the entire person or persons in question, and these facts include facts about both the past and future, since persons extend into both the past and future.

Any three-dimensionalist appeal to such future facts, however, is far less plausible given that for her, persisting objects wholly exist whenever they exist. Hence she may need to hold that multiple coincident persons would have existed even if there had been no fission. The four-dimensionalist, however, can hold that one and the same temporal part that existed prior to fission—Riker-part—in fact turns out to be part of two distinct four-dimensional persons, R_1 and R_2, *because* there is a fission event, though that same temporal part would have only been a part of one four-dimensional person if there had been no fission event. Thus the four-dimensionalist is not committed to the existence of multiple persons who are coincident (overlap) at *all* times at which they exist.

So are there two persons in existence prior to fission? Well yes and no. There is a single temporal part, Riker-part, which, as it turns out, is a part of both R_1 and R_2. So there is a sense in which two persons exist prior to fission, namely the sense in which Riker-part exists prior to fission, and Riker-part is part of both R_1 and R_2. Just as Robinson distinguishes between counting by identity and counting by constitution, Lewis approaches the issue of how many persons exist prior to fission, by distinguishing between counting by identity, and counting by identity at a time.[28] According to Lewis, persisting objects O and O^* are identical at a time t iff at t O and O^* share the same temporal part. Thus we count by identity at a time, if we count as one at t, any objects that are identical at that time. Counting by identity at a time we will say that prior to fission only one person existed, because at every time prior to fission, R_1 and R_2 are identical at those times.

Thus it is that the four-dimensionalist reconciles each of our competing intuitions about fission, and indeed our intuitions about each of the four puzzles of persistence. So what does all this tell us?

3. DIFFERENT SOLUTIONS?

We now have a fairly clear picture of how traditional versions of three- and four-dimensionalism deal with the puzzles of persistence. Looking at each of the solutions to these puzzles, it certainly appears that the two theories have quite different resources for addressing these puzzle cases, and that they respond to the puzzles in very different ways. Yet I will argue that in fact these two theories are metaphysically equivalent.

It is chapters four and five that provide the bulk of the arguments for this claim. And it is in these chapters that I begin to develop some novel accounts of persistence that, despite being broadly three- or four-dimensionalist, are not the traditional three- or four-dimensionalist theories that we have met in this chapter. In part these novel theories arise as a response to thinking about how different kinds of three- or four-dimensionalism might understand composition both at and across time, and what sort of view about simples such an understanding might bring.

Before we turn to these issues, however, recall from chapter one that in determining whether theories are equivalent, a great deal of weight is placed upon the notion of inter-theoretic translation: upon the notion of a practical translation and ultimately a correct translation. Part of the core task of chapters four and five is to show that there is a correct translation between three- and four-dimensionalism. In order to do this, however, it is first necessary to be absolutely clear what sets of sentences compose those theories, and what terminology occurs in each theory. Only once we are clear about the terms of each theory can be we begin the process of determining whether we have a practical or correct translation between the theories. Hence it is my task in chapter three to begin to define the key theoretical terms of three- and four-dimensionalism.

NOTES

[1] Leibniz (1696). pp. 133–136.

[2] Something like this puzzle is sometimes referred to as the problem of temporary intrinsics. Of course, the problem of temporary intrinsics is really just a particular subset of the problem of change: it deals with intrinsic rather than, say, relational properties. The problem of temporary intrinsics is only a more difficult problem than the problem of change as it pertains to relational properties, given certain solutions to the problem of change. We will discuss this issue later.

[3] Myro (1986); Griffin (1978); Gallois (1998).

[4] Gibbard (1975).

[5] I will often talk of objects being identical to themselves at different times, or being strictly identical across time. We might think that this does not distinguish between a

three-dimensionalist and a four-dimensionalist account, since everyone agrees that all objects are self-identical. So then doesn't everyone agree that objects are identical to themselves whenever they exist, and are strictly identical across time? Well yes, on one reading of 'strictly identical across time.' The sense I intend is the following: an object is strictly identical to itself at different times, (or strictly identical across time) just if the thing that wholly exists at one time, is strictly identical to the thing that wholly exists at some other time. Three-dimensionalist affirm this, four-dimensionalists deny it, since for the four-dimensionalist, the only thing that wholly exists at a time is a temporal part of a persisting object, and no two of these are ever identical.

[6] So three- and four-dimensionalists agree that in some sense persisting objects have a temporal dimension—they persist through time—they disagree about whether those objects are extended across that dimension, or are wholly present at each moment within that dimension.

[7] Van Inwagen (1990).

[8] Johnston (1987); Haslanger (1989).

[9] Those three-dimensionalists who defend some form of constitution relation include Wiggins (1968); Johnston (1992); Baker (1997); Thomson (1998); Doepke (1982) and Simons (1985).

[10] Baker (2000) is the one exception. She holds that constitution is the relation that holds at times, between *spatially* coincident objects.

[11] For instance Baker (2000); Thompson (1998); Doepke (1982); Simons (1985).

[12] Baker (2000) p. 33.

[13] Thomson (1998).

[14] Doepke (1982).

[15] Simons (1987). p 239.

[16] Wiggins (1980). p 70.

[17] Robinson (1985).

[18] Proponents of best candidate theories include Smart (1972) and (1973), and Nozick (1982).

[19] This also seems problematic on another front. If identity is necessary, rather than contingent, then either Riker is identical to R_1, or he is not: if there is a world in which Riker is identical to R_1, (the world where R_2 does not exist) then in the actual world Riker must be identical to R_1. You might take this either to be another reason to reject a best candidate account, or as a reason to embrace contingent identity.

[20] Parfit (1984).

[21] Robinson (1985).

[22] Perdurantists include Lewis (1983); Sider (2001); Balashov (2002); Heller (1984) and Noonan (1993).

[23] For discussion of the role of (immanent) causation between temporal parts of persisting objects see Armstrong (1980). pp 67–68; Balashov (2003b) and Shoemaker (1979).

[24] Except for spatially extended mereological simples if such do or could exist. But I take it they are the exception.

[25] Heller (1990). Chapter one.

[26] Not everyone agrees that statements of identity can ever be contingent. See for instance Kripke (1972).

I discuss contingent identity further in later chapter of the book. For now, perhaps it is easiest to think of contingent identity in terms of counterpart theory. Then 'x' and 'y', or 'Lumpl' and 'Goliath are non-rigid designators. These names have attached to them different counterpart relations, which may pick out different objects in different worlds.

[27] Though we could see four-dimensionalism as a way of making sense of Parfit's claim that it is not identity that matters in survival. It is not identity that matters insofar as objects are not strictly identical with themselves at different times: temporal parts of objects are not strictly identical, rather, persistence is a matter of the appropriate (non-identity) relation holding between temporal parts.
[28] Lewis (1983).

Chapter 3

DEFINING OUR TERMS

1. GETTING STARTED

We noted in the previous chapter that any argument to the effect that three- and four-dimensionalism are metaphysically equivalent will rely heavily on showing that the two theories are practically inter-translatable, and from there that they are correctly inter-translatable. Before we can show that there exists a practical translation, however, we must first show that there exists an assertibility mapping. That in turn requires that we have clear definitions of all of the relevant terms of the theories. Only once we know precisely what three-dimensionalists are, by their own lights, committed to, and *mutatis mutandis* for four-dimensionalists, can we show that there exists an assertibility mapping between these theories. Then, and only then can we move on to show that the theories are correctly inter-translatable.

Of course, neither the theories of three- or four-dimensionalism exist in a metaphysical vacuum. Rather, theories of persistence are closely related to both theories of time and theories of ontology. Both three- and four-dimensionalists need to decide whether to embrace a presentist or eternalist theory of time—that is, whether to hold that only the present is ontologically real (presentism), or to hold that all temporal locations are equally ontologically real (eternalism). They also need to decide which arrangements of particulars compose further objects. That is, they need to decide under which conditions composition occurs, and under which conditions it does not. There are essentially three broad views when it comes to questions of composition. The first is the view that no arrangements of concrete basic particulars ever compose some further object. This is the view that there exist only simples, and that composition never occurs. We met this view briefly in chapter one as mereological or compositional nihilism. This is not a common view, and in general is not a view I will discuss.

The second view is that for every arbitrary arrangement of concrete basic particulars, there exists some object composed of those particulars. This is the view that composition occurs under any circumstance: the view that composition is unrestricted. Finally, there is the view that only some arrangements of concrete basic particulars compose some further object—the view that composition occurs only under certain circumstances. Thus this is the view that composition is restricted.[1]

I call these additional metaphysical commitments *ancillary metaphysical commitments*. Naturally, some combinations or 'packages' of these ancillary commitments are more common than others. Typically, four-dimensionalists are both eternalists and unrestricted compositionalists, while it is less common for three-dimensionalists to adopt these particular ancillary commitments. Despite such preferences, we want to define our terms in the broadest way possible, such that the definitions are consistent with a range of ancillary commitments. In this chapter I will be concerned to show that the definitions are consistent with both eternalism and presentism, and also to show that *prima facie*, there is no reason to suppose that they are not also consistent with both restricted and unrestricted composition. This latter issue will be more fully explored in chapter four where I consider versions of three- and four-dimensionalism that have different ontological commitments. Clarifying the nature of the different packages of ancillary commitments forms a basis that allows me to argue, in the following chapters, that so long as we compare versions of three- and four-dimensionalism that adopt the same metaphysical packages, then we can show that these theories are equivalent.

I begin in section two by clarifying exactly what we might mean when we talk about the theory of four-dimensionalism, and I then proceed to clarify and define key four-dimensionalist terminology, much of which we have already encountered. I then proceed to show that these definitions are consistent with a range of ancillary metaphysical commitments. In section three I move on to consider the theory of three-dimensionalism and to define the key terms of that theory such that they are consistent with a range of ancillary commitments.

2. FOUR-DIMENSIONALISM

In general, four-dimensionalism is the thesis that persisting objects are four-dimensional. Of course, the strength of this claim might vary. At one end of the spectrum it might be the claim that some, or all, actual persisting objects are four-dimensional, while at the other end of the spectrum it might be the claim that all possible persisting objects are four-dimensional. The same holds true *mutatis mutandis* for the claims a three-dimensionalist might make. For our purposes, it does not matter what we take the strength of the claims to be, so long as we are considering versions of each theory that are making claims of the same strength. (The view that every actual persisting object is four-dimensional, cannot be equivalent to the view that every possible persisting object is three-dimensional). Given the sorts of arguments mustered in favour of three- or four-dimensionalism, it is plausible that proponents of each view see themselves as providing an account of the nature of persistence in general, rather than merely an account of the persistence of actual objects. So I will define the scope of three- and four-dimensionalism accordingly, in terms of them being accounts of what it is for any *possible* object to persist. But nothing hangs on this choice. Those who think that three- or four-dimensionalism might be merely contingent claims about actual objects can feel free to supplement the appropriate definition of each thesis. Thus:

> 4D: Four-dimensionalism is the thesis that every possible persisting object is four-dimensional.

What is it for an object to be four-dimensional? That persisting objects are four-dimensional is most usually understood as the claim that persisting objects perdure—that they persist by being composed of a series of distinct temporal parts. This is the view I have referred to as perdurantism. However, the view that persisting objects are four-dimensional does not entail that they are composed of temporal parts, any more than the view that objects are three-dimensionally spatially extended entails that they are composed of spatial parts. In each case this is the more common view, but it does not follow from the notion of spatial or temporal extension. Thus for future clarity, it will serve us well to define the general idea of four-dimensionality in a way that does not depend on the notion of a temporal part. Thus we have the following:

> A persisting object *O* is four-dimensional iff *O* is temporally extended and it is not the case that *O* wholly exists at any temporal location at which it exists.

Now, the notion of 'wholly existing' seems to require further definition, and this is a matter we will return to in chapters five and six. What does it mean to fail to wholly exist at a time? Given perdurantism, four-dimensional objects exist at a time in virtue of some temporal part existing at a time, hence perduring objects only *partly* exist at any time—so they fail to wholly exist whenever they exist.

Let us consider perdurantism for a moment before returning to the issue of what it might mean for a four-dimensional object to fail to wholly exist at a time if it does not simply mean that the object in question perdures.

Perdurantism avails itself of the notion of a temporal part. What exactly is a temporal part? There are a number of ways of defining 'temporal part'. For some, a temporal part, *TP*, of a four-dimensional whole, *O*, is an object that overlaps all of *O's* spatial parts at the time or times at which *TP* exists. So instantaneous temporal parts are three-dimensional slices of four-dimensional wholes, whilst extended temporal parts are fusions of temporally contiguous three-dimensional slices. Ted Sider, for instance, defines temporal parts thus. This sense of temporal part is the one that I have hitherto been employing. For others though, a temporal part need not overlap *every* spatial part of a four-dimensional whole at the time or times at which that temporal part exists. Rather, a temporal part is an object that is wholly overlapped by *some* part of the four-dimensional whole at all times at which that temporal part exists.

To discriminate between these two senses of temporal part, let us distinguish what I will call a *maximal temporal part* and a *non-maximal temporal part.* A maximal temporal part is an object that overlaps every spatial part of the four-dimensional whole of which it is a part, at the time or times at which the temporal part exists. Thus amending slightly Ted Slider's definition of a instantaneous temporal part:[2]

> x is an instantaneous maximal temporal part of y at instant $t =_{\text{df}} 1)$ x is part of y and 2) x exists at, but only at t and 3) x overlaps every part of y that exists at t.

Instantaneous maximal temporal parts are thus also sometimes referred to as temporal slices, since they are three-dimensional slices of four-dimensional wholes. Of course, not all temporal parts are instantaneous: some have an extended duration and are thus extended maximal temporal parts. Amending slightly Ted Sider's[3] definition of an extended temporal part we can define an extended maximal temporal part as follows:

> An extended maximal temporal part of y during temporal interval T is an object that exists at all and only times in T, is part of y at every time during T and at every moment in T overlaps everything that is part of y at that moment.

An extended maximal temporal part is an object that exists only during a particular temporal interval, and which during that interval has the spatial dimensions of the object of which it is a temporal part. Extended temporal parts are sometimes also referred to as temporal stages, since, in effect, they are stages of the object of which they are parts.

Temporal parts that do not overlap all of the spatial parts of the four-dimensional whole of which they are parts are *non-maximal* temporal parts. We can define an instantaneous non-maximal temporal part as follows:

> x is an instantaneous non-maximal temporal part of y at instant $t =_{df}$ 1) x is part of y and 2) x exists at, and only at t and 3) some part of y that exists at t overlaps every part of x.

Then an extended non-maximal temporal part will be defined as follows:

> An extended non-maximal temporal part of y during T is an object that exists at all and only times in T, is part of y at every time during T, and at every moment in T overlaps some part of y at that moment.

When I use the locution 'temporal part' I intend this as shorthand for 'maximal temporal part'. Where I intend to talk of non-maximal temporal parts I will specify this on each occasion.

With these definitions in mind, we can now define perdurance as follows:

> P: An object O perdures iff it is a mereological fusion of temporal parts.

This is what we might think of as a weak definition of perdurance. Given P, all that is required for an object to perdure, and hence persist, is for that object to be a fusion of temporal parts. There are those who argue, however, that mere mereological fusion is insufficient for persistence. Armstrong asks to imagine a case where we discover that what we thought was a persisting object—a dog, say—is in fact a series of distinct qualitative dog-duplicates: the 'dog' is destroyed every second and replaced by a qualitative duplicate.[4] Armstrong's intuition is that in this case we do not have a persisting dog, because the appropriate causal connections between each dog temporal part fail to exist. Yet this object does perdure, suggesting that perdurance is not a sufficient account of persistence.

There is a distinction to be drawn between a normal persisting dog that has the normal causal connections between each of its maximal temporal parts, and the case described above. The point is that there is some salient difference between some gerrymandered fusion of temporal parts, and a fusion of temporal parts that are causally connected in certain ways. We could conclude from this that perdurance is not sufficient for true persistence: in order to persist, objects must not only perdure, but their temporal parts must be appropriately causally connected. Thus the world is full of perduring objects, some of which persist and some of which do not. On the other hand, we could hold that perdurance is sufficient for persistence and hence the destroyed and replaced 'dog' considered above does persist, but perhaps that object is not really a dog. For perhaps *dogs* only persist if their temporal parts are appropriately causally connected.

Ultimately this is a mere semantic debate: do we concede that gerrymandered objects *persist* but are saliently different from everyday objects because they lack the appropriate causal connections, and thus we say that the odd gerrymandered 'dog' persists but is not really a dog at all (but a mere persisting mereological fusion of dog-like parts), or do we say that only the objects whose temporal parts are appropriately connected persist, and that other perduring objects are merely temporally extended non-persisting objects? I opt for the simpler view that perdurance is sufficient for persistence, but that causal connections may be necessary for the persistence of particular kinds of objects, such as dogs. But nothing hangs on this way of expressing the distinction.

Given this, we will say that perdurantism is the view that:

PER: Every possible persisting object perdures.

Thus if perdurantism is true then four-dimensionalism is true, since if objects perdure then they are temporally extended and they do not wholly exist at any time at which they exist. The reverse is not the case. Perdurantism is only one way that four-dimensionalism could be true. What is the alternative to perdurantism? This is an issue that I will discuss at greater length in chapter five. But the intuitive idea is straightforward. The idea of a four-dimensional non-perduring object is analogous to the idea of a spatially extended object that lacks spatial parts. A spatially extended mereological simple exists at multiple spatial locations, but not in virtue of having some spatial part that exists at those locations. Yet the simple does not 'wholly exist' at any of those locations.[5] The simple is, we might say, 'spread out' across space—only 'some' of it exists at each location despite the fact that there is no *part* that exists at those locations. The same will be true of four-dimensional objects that lack temporal parts: such objects are temporally simple (they lack temporal parts) just as our spatially extended object is spatially simple (it lacks spatial parts) and yet the four-dimensional object is still 'spread out' in time and thus does not wholly exist at any temporal location.

Such temporally extended temporal simples do not persist by perduring. Let us call the manner in which they persist *terdurance*. Then:

T: An object *O* terdures just if it exists at multiple temporal locations in virtue of being a temporally extended temporal simple that does not wholly exist at any of the temporal locations at which it exists.

Then *terdurantism* is the view that:

TER: Every possible persisting object terdures.

Then four-dimensionalism is true if either perdurantism or terdurantism is true.

Now that we have clarified what it would be for four-dimensionalism to be true, and have further defined the key terms of the theory, we can move on to confirm that these definitions are consistent with a range of additional metaphysical commitments.

2.1 Ancillary Commitments and Consistency

Earlier in this chapter I outlined four different metaphysical views—two sets of contraries—that one might combine with four-dimensionalism. The first of these pairs is presentism and eternalism. Is four-dimensionalism as defined consistent with both presentism and eternalism? Four-dimensionalism is the thesis that persisting objects are temporally extended and do not wholly exist at any temporal location. Most four-dimensionalists (in the form of perdurantists) are eternalists. That is hardly surprising. Eternalism is the thesis that all temporal locations are equally ontological real, so the combination of eternalism and four-dimensionalism is a natural one—it is, essentially, the block universe view. So both perdurantists and terdurantists should have no difficulty in embracing eternalism.

But what of presentism? Berit Brogard argues that four-dimensionalism in the form of perdurantism is not only compatible with presentism, but that such a combination has much to recommend it.[6] Her idea is that perduring objects are composed of temporal parts, but that at any time t, the only part that exists is the t-part. Is definition P consistent with presentism? That depends on how we construe the notion of mereological fusion. We might think that if there exists a fusion of objects that exist at different times, then each of those times must be equally ontologically real, for a fusion just is an object that tenselessly has as parts, each of the particulars that it fuses. But it is conceivable for the presentist to read 'fusion' such that there can exist a fusion of objects that did exist, do exist and will exist. Then a perduring object is an object composed of temporal parts some of which did exist, one of which does exist, and some of which will exist. Those who hold that 'fusion' cannot be read this way can amend the definition such that:

> P1: An object x perdures iff (i) it has temporal parts and (ii) at any given time t, only the t-part exists.[7]

If P is not consistent with presentism, then P1 is. So we have a definition of perdurantism that is consistent with both eternalism and presentism.

What of four-dimensionalism construed more broadly? The combination of non-perdurantist four-dimensionalism (terdurantism) with presentism is a stretch. One can be forgiven for being perplexed about what it means to say that only one temporal location—the present—exists, and that there exists a temporally extended temporal simple that does not wholly exist at any temporal location, and thus does not wholly exist in the present. The presentist perdurantist can at least make sense of why four-dimensional objects do not wholly exist at any temporal location, by appeal to the fact that they are fusions of temporal parts some of which did, do, and will exist. Presentist terdurantism is, I suppose, analogous to the view that only the spatial point 'here' is ontologically real—there is some absolute 'here' just as there is some absolute 'now' for the presentist—and that despite this, there exist spatially extended mereological simples that are not wholly located at any one spatial point.

While I concede that both of these views (whether in their spatial or temporal guise) are peculiar, I do not see that they can be ruled out *a priori*, and thus I do not see that terdurantism is actually inconsistent with presentism. The terdurantist presentist must simply hold that persisting objects are temporal simples that are temporally extended insofar as there are multiple temporal locations at which they fail to wholly exist. Odd, but not inconsistent.

So four-dimensionalism is consistent with both presentism and eternalism. What of the other pair of metaphysical commitments: restricted and unrestricted composition? It seems that our definitions are consistent with both these views. For our definitions tell us nothing about under what circumstances composition occurs. They do not tell us which arrangements of particulars at a time compose some instantaneous object, nor which combinations of instantaneous objects compose persisting objects. All they tell us is that if there are actual or possible persisting objects, then those object are four-dimensional. They do not tell us anything about *which* objects exist. So, *prima facie* at least, there is no reason to suppose that being committed to four-dimensionalism as defined, rules in or out being committed to restricted or unrestricted composition. Thus as we intended, our definitions are consistent with a range of ancillary metaphysical commitments.

In section three and four we will define the terms of the three-dimensionalist theory. Before that, it is worth briefly clarifying the relation between traditional perdurantist four-dimensionalism and the view that Ted Sider defends—the stage view.

2.2 The Stage View

In general when I talk of perdurantism, I refer to the view that persisting objects are four-dimensional and exist at multiple times in virtue of having temporal parts that exist at those times. Construed like that, perdurantism incorporates another variant of four-dimensionalism: the stage view. According to most perdurantists, the everyday persisting objects of our ontology—continuants, as they are often known—should be identified with the mereological sum of temporal parts, that is, with four-dimensional wholes or space-time worms. This view is thus often known as the *worm view*. On this view a term such as 'dog' refers to the fusion of maximally dog-related temporal parts—the maximal four-dimensional space-time worm each of whose temporal parts are related by the dog-preserving relation.

According to the stage view though, continuants should be identified with a temporal part or temporal stage.[8] Thus our term 'dog' refers to a temporal part of the maximally dog-related four-dimensional worm. The stage view then, countenances the existence of the same four-dimensional objects united by the same similarity cum gen-identity relations as does the worm view. But rather than holding that our terms refer to those four-dimensional wholes, and thus that continuants persist in virtue of having different temporal parts at different times, the stage view instead introduces the notion of a temporal counterpart relation in order to explain how continuants persist. The temporal counterpart relation is the same unity or gen-identity relation that the worm theorist appeals to in unifying the succession of temporal parts that compose a four-dimensional whole. The stage theorist appeals to the temporal counterpart relation to analyse statements about the past and future of continuants. Thus statements about the past of some continuant *C* are true just if the temporal part that is the referent of '*C*' has some past temporal counterpart of which the statement was true. Hence 'my dog was dirty' is true just if my dog—identified as the current temporal part of the dog-related space-time worm—has some past temporal counterpart that has the property of being dirty.

In essence then, the debate between the worm and the stage view is a debate about the reference of our ordinary terms, not a substantial debate about ontology. Strictly speaking we might say that on the stage view ordinary objects do not perdure, but rather, persist by having the relevant temporal counterparts. This is, however, a minor variation on traditional perdurantism, and one that will make little difference when it comes to the project of determining whether three and four-dimensionalism are metaphysically equivalent. In general I will include both the worm and the stage view under the umbrella of 'perdurantism' except in relevant instances where I specify otherwise.

We can now turn to define the terms of the theory of three-dimensionalism.

3. THREE-DIMENSIONALISM

Three-dimensionalism is the view that objects persist through time by being wholly present at every time at which they exist, that is by enduring through time. What does it mean to say that an object endures through time, or that it is wholly present at every time at which it exists? These questions are not easy to answer. Ted Sider, for instance, holds that there is no coherent way to define endurance such that the definition will hold for composite objects.[9] And Trenton Merricks argues that the only plausible way to define endurance entails that presentism is true.[10] In the following section I will consider some of the problems associated with defining endurance, before in the final section of this chapter moving on to develop a novel definition that is compatible with a number of ancillary metaphysical commitments.

3.1 Definition Difficulties

In keeping with the definition of four-dimensionalism that we adopted, let us say the following:

> 3D: Three-dimensionalism is the thesis that every possible persisting object is three-dimensional.

What it is to be three-dimensional is to fail to be temporally extended, that is, to have only three spatial dimensions. Since three-dimensional objects persist by enduring, we could say that three-dimensionalism is the thesis that every possible persisting object endures. Then the question is, what is it to endure?

Typically, three-dimensionalists talk of enduring objects being wholly present at every time at which they exist, and being strictly identical to themselves at each of those times.[11] This suggests that we define endurance in terms of identity:

> ID: An object O endures through interval T iff for any two times t and t' contained in T, O at t is identical to O at t'.

Unfortunately, as Merricks notes, defining endurance in terms of identity is problematic.[12] A definition of endurance ought clearly to distinguish it from perdurance (and terdurance). We can see how this is supposed to work with ID. The perdurantist holds that at each time at which a persisting object O exists, it is a distinct part of O that exists at that time. Thus if O persists through interval T, and t_1 and t_2 are instants in T, then O-at-t_1 is distinct from O-at-t_2. Thus if O perdures it does not meet definition ID.

The problem with ID is that it can be read in a way that is compatible with perdurantism. For the perdurantist (who holds the worm view) holds that '*O*' refers to the entire four-dimensional object. So when the perdurantist points to *O* at t_1, and says '*O*', he intends the reference of '*O*' to be the entire four-dimensional object, and not the temporal part *O-at-*t_1. The four-dimensional object *O*, which we refer to at t_1 by pointing to part of *O*, is at all times self identical. Understood in this sense, it is true that *O* at *t* is identical to *O* at *t*'. The problem with ID is that without first providing an account of what it is for *O* to exist at a time, ID does not rule out perdurance. Nor, for analogous reasons, does it rule out terdurance.

So perhaps as an alternative to ID, we could define endurance as follows:

> S: An object *endures* iff it persists and has only spatial extension.

The problem here, as Merricks points out, is that to say that an object has only spatial extension is to say that while not all of an object's parts exist at one point in space, all of its parts exist at one time.[13] Thus to say that an object is not temporally extended is to say that all of that object exists at a time, and that seems to be no more than to say that it is wholly present at every time at which it exists. Ultimately then, to define endurance we need to be able to define what it is to exist at a time, and what it is to exist at a time for the three-dimensionalist, is to be wholly present at that time. This would suggest that we say the following:

> E: An object *endures* iff it exists at multiple times and is wholly present at each of those times.

We then have the notion of 'wholly present' left undefined. Ned Markosian suggests the following:[14]

> WP: An object *x* is *wholly present* at *t* just in case (a) *x* exists at *t* and (b) it is not the case that there is a *y* such that *y* is a temporal part of *x* at some time other than *t*.

Combining E and WP we could then define endurance as follows:

> E1: An object *endures* iff it exists at multiples times and is wholly present at each of those times, where an object is wholly present at a time iff (a) it exists at *t* and (b) it is not the case that there is a *y* such that *y* is a temporal part of *x* at some time other than *t*.

The difficulty with this definition is that an object endures just if it is not the case that it perdures. This is problematic for three reasons. First, three-dimensionalism is the thesis that all possible persisting objects endure. By substituting in E1, three-dimensionalism would be the thesis that no possible objects perdure. But three-dimensionalism is not just the claim that perdurantism is false, rather, it is the more general view that persisting objects are three-dimensional and not four-dimensional. Three-dimensionalists hold that persisting objects *neither* perdure *nor* terdure. Yet E1 leaves it open that a four-dimensional non-perduring object endures, or conversely, that an enduring object could be four-dimensional.

Even if we added in a further clause to E1 in order to rule it out as possible that a four-dimensional object could count as enduring, there are further problems for this definition. Some three-dimensionalists hold that the very notion of a temporal part is incoherent,[15] or that although in some abstract sense they understand the notion of a temporal part, they can make no sense of the idea that objects are composed of temporal parts.[16] One who holds either of these views, justly or not, would not want to define endurance in terms of the absence of temporal parts.

A broader problem with this approach is that three-dimensionalism is typically thought to be the more intuitive, 'folk friendly' view that significantly pre-dates four-dimensionalism. But if the *only* coherent definition of endurance is in terms of the negation of four-dimensionalism in the form of a negation of both perdurance and terdurance, then this significantly undercuts the idea that there was any clear, coherent view about persistence prior to a four-dimensionalist account being constructed. That is not decisive. Still, all things considered it would be better if the three-dimensionalist could state her thesis in a way that does not involve negating the thesis of four-dimensionalism. So let us put aside E1.

The idea that an object lacks temporal extension if not all of its parts exist at one point in *space*, but all of its parts do exist at one *time*, suggests another way to define the notion of being wholly present. The idea of being wholly present is the idea that *all* of an object is present. Implicitly then, the notion of being wholly present invokes the notion of parthood: for surely if all of an object is present, then all of its parts must be present. This suggests the following definition of 'wholly present':

> WP1: An object x is *wholly present* at a time just in case all of its parts are present at that time.

Both Sider and Merricks consider this definition of 'wholly present'. Sider rejects WP1 on the grounds that it fails for any object that loses or gains parts over the course of its existence. Given what we know about microphysics, it follows that no composite object would ever be wholly present, and thus would not endure. Merricks draws the somewhat different conclusion that this shows that the three-dimensionalist must accept presentism. For if presentism is true, then all of an object's parts are present at every time at which it exists: it only exists in the present, and in the present, all of its parts are present. It was, however, our goal to define three-dimensionalism such that it was consistent with a range of metaphysical views, including eternalism. Given this, WP1 is unsuccessful since three-dimensionalists *do* hold that everyday composite objects endure. We are faced, then, with the prospect of being unable to define endurance either in terms of identity, spatial extension, or parthood. Recently, however, Sider has made a further suggestion, and it is to this that we will now turn.

3.2 Possibilist Suggestions

Consider the following definition suggested by Sider:

> WP2: x is *strongly wholly present* throughout interval T iff everything that is at any time in T part of x, exists and is part of x at every time in T.[17]

This definition of 'strongly wholly present' bares a striking similarity to WP1. If we take the temporal interval T to include the entire lifespan of x, then WP2 is equivalent to WP1. In both cases x is (strongly) wholly present at all times in its life just in case at all times at which x exists, it has all and only the same parts at each of those times. Thus if we were to define endurance as follows:

> E2: x *endures* iff x exists at multiple times and x is strongly wholly present at all of those times.

then, again, the only objects that endure are objects that either have no spatial parts, (simples) or which neither gain nor lose spatial parts (mereological constants). In the actual world, then, we would be forced to conclude that only simples endure. Since most three-dimensionalists do not hold the view that only simples endure, this leads Sider to suggest E3:

> E3. An object x *endures* iff x exists at multiple times and it is possible that x is strongly wholly present for some interval of time T.[18]

Given E3, if an object endures then it follows that the object does not perdure in the traditional sense of having, at each instant at which it exists, some instantaneous temporal part. Since T is a temporal interval, it follows that if x is strongly wholly present through T, then x is not composed of any *instantaneous* temporal parts during T. For suppose t and t' are temporal instants within T. Suppose that at t there is some instantaneous temporal part of x. By definition, this temporal part exists at t, and at no other temporal instant. Then there is some part of x that exists in T and which is not part of x at all times in T: that instantaneous temporal part. So it follows that if x is strongly wholly present through T, no such temporal part exists during T. So if an object endures according to E3, then that object is not composed of instantaneous temporal parts.

Still, E3 is not inconsistent with the idea that objects are four-dimensional and perdure, but are the fusions of extended temporal parts rather than instantaneous ones. Interval T might be shorter than the temporal extent of the shortest extended temporal part of any perduring object, in which case that object would, according to E3, endure (as well as perdure). Since the three-dimensionalist does not just want to rule out the existence of instantaneous temporal parts, she wants to show that objects are in fact wholly present at every moment at which they exist, it follows that E3 is too weak. Moreover, not only does E3 not rule out the existence of *some* perduring objects counting as enduring, it also does not rule out that *any* terduring objects could count as enduring. For I see no reason why a four-dimensional object that lacks temporal parts could not *actually* be strongly wholly present through some interval, much less why it should not be *possibly* strongly wholly present through some interval.

Even if we develop a stronger version of this type of definition, we will still not succeed. Consider E4. While E4 does rule out that any object that endures also perdures, it does not rule out that any object that endures is not also four-dimensional.

> E4: An object x *endures* in a world w just in case x exists at multiple times in w and there is some world z such that in z, x is strongly wholly present throughout interval T, where interval T includes every time t during which x exists in z.

E3 and E4 share the same underlying problem: the appeal to possibility. Each tells us that some object endures just in case it is *possible* that it persists in some way, rather than in virtue of how it *actually* persists, and this surely fails to capture something important in the three-dimensionalist's notion of 'wholly present'.

The three-dimensionalist began by claiming that objects persist by being wholly present whenever they exist. But almost all of the objects of our day-to-day ontology fail to be actually strongly wholly present throughout even a short interval. If the three-dimensionalist accepts E4, she ought more properly to say that objects endure by having the property of being-possibly-strongly-wholly-present-throughout-their-existence. But this seems to tell us little about the way objects actually persist, other than the fact that it is possible that they do not perdure.

E4 does capture something of the three-dimensionalist intuition. It captures the intuition that it is possible that enduring objects have identity of parts over time. In a world where *O* is strongly wholly present across time, every part of *O* at *t*, is identical to every part of *O* at *t*'. This, of course, is precisely what perdurantism rules out. Still, here we have a conceptual claim about what is possible, not about what is actual. Indeed, matters are more dire than this. Even if we put aside the worry about non-perduring four-dimensional objects counting as enduring on this definition, E4 still does not succeed. For suppose that *w* is the actual world. Given E4, *O* endures *w* just if there is some world *z* such that *O* is strongly wholly present in *z*. If *O* is strongly wholly present in *z*, then *O* does not perdure in *z*. This means that *O* endures in the actual world just in case there is some possible world *z* in which *O* does not perdure.

Actually, that is not quite right. It could be that there is some world *z* in which objects are strongly wholly present, and yet none of the objects in *z* might have counterparts of, or be trans-world identical to, any actual objects. Then it would not follow from the existence of *z*, that the actual world is an enduring world. If we think though, that for any actual object *O* there *does* exist some strongly wholly present counterpart of *O*, then we think that the actual world is an enduring world just in case perdurantism is not necessarily true. This would be extremely problematic since definitions E3 and E4 would then be consistent with it being the case that actual object *O* counts as enduring despite being composed of a series of distinct temporal parts. Yet surely the fact that *O's* counterpart is strongly wholly present in world *z*, tells us nothing about *O* in the actual world.

Earlier I defined three-dimensionalism as the thesis that every possible persisting object is three-dimensional, and four-dimensionalism as the thesis that every possible persisting object is four-dimensional. Thus three-dimensionalism is also the thesis that every possible persisting object endures. Thus it is the view that every possible world is what we might call an *enduring world*: a world in which every persisting object endures. Perdurantism, then, is the thesis that every world is a *perduring world*: a world in which every persisting object perdures. Then suppose we think that for any actual object *O*, there exists some strongly wholly present counterpart of *O*. Then perdurantism would turn out to be false, though the falsity of perdurantism would not entail the truth of three-dimensionalism—since it would not entail that there are *no* worlds in which objects terdure or perdure—but worse still, although it *would* entail that *actual* objects endure, it would not entail that they do not also have temporal parts—that they perdure—or that they are not four-dimensional—that they terdure. (This is even more puzzling for those who think that three- and four-dimensionalism are contingent theories about the manner in which actual objects persist. In that case both three and four-dimensionalism might turn out to be true of the actual world: four-dimensionalism might be true because actual objects persist in virtue of being composed of temporal parts, and three-dimensionalism might be true because actual objects endure in virtue of having counterparts that are strongly wholly present over some interval of time.)

Of course, just as we might reject the sort of flexible counterpart relation that allows that there is a world in which my counterpart is a hunk of Ashgrove cheese, so too we might reject the claim that any actual perduring object could have a strongly wholly present counterpart (and thus count as enduring). And perhaps there are reasons to reject these claims. E4 requires that some actual object *O* endures only if there is some counterpart of *O* that is strongly wholly present, and thus mereologically constant, throughout its existence. It is not at all clear though, that it is even logically possible that actual objects have counterparts that are mereologically constant: perhaps it is an essential property of some or all actual composite objects that they change over time. In that case there is reason to suppose that no actual composite object has a counterpart that is strongly wholly present.

If that is so, then it is good reason to reject E4. For although the three-dimensionalist is committed to the idea that an object considered at one time, is strictly identical to that object considered at another time, it is unclear why she should be committed to the idea that it is possible that enduring objects be mereologically constant. If it is impossible that *any* actual object have a strongly wholly present counterpart, then it is certainly impossible that any actual *perduring* object has a strongly wholly present counterpart. However, it then turns out to be impossible that *any* actual object endures, so E4 must be abandoned.

E3 fares better on this score. For E3 requires only that some actual object *O* endures if there is a counterpart of *O* that is strongly wholly present through some interval of time *T*. Since interval T could be of very short duration, it is not obvious that any actual object *O* should fail to have a counterpart that is strongly wholly present through *some* interval *T*. Since it is plausible that actual objects do have counterparts that are strongly wholly present through some interval *T*, however, we are brought back to our original problem that an actual object *O* can endure solely in virtue of the existence of such a counterpart, regardless of the manner in which *O* persists in the actual world. So we should also reject E3.

4. A NEW DEFINITION OF ENDURANCE

We have seen that attempting adequately to define endurance seems ultimately to return us to the issue of parthood. The problem is that there seems to be no way for the three-dimensionalist to say both that all of an object's parts exist whenever it exists (it is wholly present at all of those times) and that an object's parts change across time. Yet there is something familiar about this tension: it seems to mirror an earlier problem we encountered with change—the problem of how one and the same object can change properties over time and yet be strictly identical with itself whenever it exists.

Perhaps reconsidering the three-dimensionalist's responses to this problem may provide insight into the analogous problem of parthood across time. In the following section I argue that there are some problems with the three-dimensionalist accounts of property instantiation. In particular, although adverbialism is able to embrace the idea that properties are instantiated *simpliciter*, it seems unable to explicate why it is that at some times properties are manifest, and at other times are not manifest. Consideration of these issues leads me to argue that the idea of possessing a property *simpliciter* is not univocal. There are at least two separate notions at work: a strictly metaphysical notion and a hybrid metaphysical/semantic notion. By explicating these two notions, it will be possible to see how the three-dimensionalist can both embrace a metaphysics of underlying non-relational properties that objects can possess at times, while also explaining why it is that there are times at which these properties are manifest and times at which they are not. Then in section 4.2 the mechanics that underlie this account of property instantiation will be put to work in the arena of parthood. This will allow us to formulate a definition of endurance.

4.1 Instantiating Properties *Simpliciter*

Recall that one of the three-dimensionalist responses to the problem of change is indexicalism, which relativises properties to times such that if object *O* is red at *t*, then it has the property of being red-at-*t*. Indexicalism, however, is problematic on two fronts. A common objection is that given indexicalism, no properties are ever really intrinsic, rather, what appear to be intrinsic properties are disguised relations to times.[19] This is the problem of temporary intrinsics. And there is a second problem. Consider a ball that is red at t_1, t_2 and t_4, and blue at t_3. We want it to be the case that there is some sense in which the ball is red at t_1 and t_2 in virtue of instantiating the same property at each of those times, whether this amounts to instantiating the same universal of redness at each time, or to having an instance of redness that persists. But on this view although the ball has all the same properties at each of the times at which it exists, (this is why it is strictly identical across time) the property the ball has at t_1 in virtue of which it is red at t_1, is the property red-at-t_1, which is a distinct property to the property the ball has a t_2 in virtue of which it is red at t_2—red-at-t_2. At each time at which the ball is red, it is so in virtue of instantiating a *different* property.[20]

Considerations such as these led Johnston to develop adverbialism, according to which the *instantiation* of properties is temporally modified.[21] Thus the adverbialist can maintain that the ball really does instantiate the very same property—redness—at each time at which it is red, the redness is merely instantiated in different temporally modified ways: it is had $t_1$ly, $t_2$ly and $t_4$ly. So properties are instantiated *simpliciter*, it is merely that they are instantiated in different ways at different times.

This means, however, that the adverbialist is committed to the ball being red *simpliciter* at every time at which it exists. That is, adverbialists are committed to the ball possessing redness at t_3 (when the ball is manifestly blue) although at t_3 that redness is not possessed $t_3$ly. It is not clear that all those who write in this area see this,[22] but it must be so. Certainly the ball does not possess redness $t_3$ly, but at t_3 it must possess redness. For if it were not the case that the ball possesses redness *simpliciter* at t_3, then there would be an interval—t_1 to t_2—over which the ball has redness *simpliciter*, and a moment—t_3—at which it does not. Assuming that redness *simpliciter* is an intrinsic property, we would have a return of the problem of temporary intrinsics. The price of both solving the problem of temporary intrinsics *and* having an underlying ontology of properties that are possessed *simpliciter* but are had in varying ways, is that the underlying properties remain *possessed* when they are not *expressed*. The analogy here[23] is with modal properties under the assumption of strict trans-world identity. All super-models have the property of being fat $w$ly, (where w is some world in which those models are fat). But even though the models have the property of being fat $w$ly, they do not manifest fatness in the actual world, because the fatness property is not instantiated in the actually manner. The fatness property, like the redness property at t_3, is possessed but not manifested.

Call this sense of instantiating a property *simpliciter* the M-*simpliciter* (for metaphysically *simpliciter*) sense. This is a strictly metaphysical notion, since it is the sense in which there is a non-relational, non-indexical property that plays a crucial constant role in explaining property attributions at times. It is the sense in which the ball instantiates the same property—redness—at t_1, t_2 and t_4.. If an object possesses some property P M-*simpliciter*, then it possesses *P* at every time at which it exists, regardless of whether *P* is manifest at that time.

Notice then, that we cannot explain the ball's not being manifestly red at t_3 in terms of it failing to possess redness (failing to be red M-*simpliciter*). This is where the second notion of instantiating a property *simpliciter* comes into play. The second notion has both a metaphysical and a semantic component. The metaphysical component addresses the issue of why it is that at certain times certain properties are manifest and at other times not manifest, given that the underlying properties that are possessed M-*simpliciter* are possessed at every time. For we surely want to say that whenever the ball is manifestly red, it is just plain red: it is red *simpliciter*. Moreover, we want there to be something in common between all and only the times when the ball is manifestly red. One way of capturing this idea is in terms of a semantic component to this notion, which addresses the issue of what is semantically in common between utterances, made at different times, which appear to attribute the same property. Thus it is the issue of what is in common between an utterance of 'the ball is red' made at different times. Call this sense of simpliciter the S-*simpliciter* sense (for semantically simpliciter).

The notion of instantiating a property S-*simpliciter* is the notion of simply being manifestly red. So if an object is red M-*simpliciter* but not red $t_3$ly, then at t_3 it is not red S-*simpliciter* since an observer in ideal circumstances would judge that the object is not red. So the notion of having a property S-*simpliciter* cannot simply be the notion of having a property M-*simpliciter*. For if an object ever instantiates a property M-*simpliciter*, then it always instantiates it whether it is manifest or not, and hence whether it is possessed S-*simpliciter* or not.

An obvious suggestion would be to hold that what explains the manifestation or not of redness at different times, are the various second-order properties of redness being instantiated in different temporally modified ways at different times. Suppose we are considering the ball's manifest redness at t_1. Then we might try conjoining the property of being red M-*simpliciter* with one of the second-order properties of instantiating redness in a particular way. We might say that the ball is red S-*simpliciter* at t_1, in virtue of being red M-*simpliciter* and having the second-order property of being red $t_1$ly.

There is a problem here though. If the property of redness is possessed $t_1$ly, then there is a second-order tenseless property of instantiating the property of redness $t_1$ly—and similarly for the property of instantiating the property of blueness $t_3$ly and so forth for all of the ball's properties. Why is the adverbialist committed to such a family of properties? Why can she not hold that there is only one property—redness—and it is simply a tensed fact that redness is instantiated in different ways at different times? In that case there is no intrinsic property of instantiating redness $t_1$ly, say, in addition to instantiating redness.

One reason is that we might think that as a matter of logical truth, a second-order property is instantiated whenever a property is instantiated in a particular way. When the ball is red at t_1, its redness is instantiated $t_1$ly, and thus it has the second-order property of having a certain property instantiated in a particular way: the property of having the property of redness instantiated $t_1$ly.

Moreover, if instantiating redness in some temporal manner is a genuine way of being red, then it is presumably not a mere relation to a time (this would in any case just get us back to indexicalism) but something more substantial. This thought would lead to a version of the argument from temporary intrinsics to the effect that if at t_1 the ball has the property of having redness instantiated $t_1$ly, and if at t_3 it fails to have that property, then since the ball is strictly identical across time it must have contradictory properties: it both has, and fails to have, the property of redness being instantiated $t_1$ly. Thus at all times at which it exists the ball has the property of instantiating redness $t_1$ly.[24]

But then consider the truth condition we just considered for being red S-*simpliciter*. We said that the ball is red S-*simpliciter* at t_1 in virtue of being red M-*simpliciter* and having the second-order property of being red $t_1$ly. That, however, will not do. For this condition remains true at t_2: at t_2 the ball is red M-*simpliciter* and is red $t_1$ly. Yet the ball is not red S-*simpliciter* at t_2 in virtue of being red M-*simpliciter* and being red $t_1$ly. This is even more apparent at t_3, when the ball is red M-*simpliciter* and is red $t_1$ly, yet is clearly not red S-*simpliciter* since we would judge that it is manifestly blue.

So the three-dimensionalist cannot think that being red S-*simpliciter* is either a simple first-order property, or even a univocal second-order property, since at every time t_n, being red S-*simpliciter* depends on having some different second-order property of possessing redness $t_n$ly at different times t_n. So on each occasion that the ball is red, it is so in virtue of instantiating a different second-order property: redness $t_1$ly at t_1, redness $t_2$ly at t_2 and so forth.

Another option for the three-dimensionalist is to say that being red S-*simpliciter* is a *complex* conjoined second-order property. For the ball has the second-order complex conjoined property of having the properties of redness $t_1$ly and $t_2$ly and $t_4$ly. Then the truth conditions for an utterance of 'the ball is manifestly red' might be:

(a) the ball is red M-*simpliciter*

and

(b) the ball has the second-order properties of being red $t_1$ly and red $t_2$ly and red $t_4$ly.

Call the conjoined property of being red M-*simpliciter* and having the relevant second-order properties the *R* property. Then we could say that the ball is red S-*simpliciter* at *t* just if it has the *R* property at *t*. And clearly the ball has the univocal *R* property whenever it is right to judge that it is manifestly red. So each time the ball is red S-*simpliciter*, it is so in virtue of instantiating the same property—*R*—at each of those times.

This will not do. First, the sense in which the *R* property is the property *in virtue* of which the ball is red S-*simpliciter* at, say t_1, is thoroughly derivative on just one of the conjuncts: the conjunct that specifies that the ball is red $t_1$ly. So too *mutatis mutandis* for every other time at which the ball is red. Worse still, what holds true for the property of being red M-*simpliciter* also holds true for the *R* property. Just as the ball is red M-*simpliciter* at t_3, so too the ball instantiates the *R* property at t_3, (as it must if it is strictly identical across time). But if instantiating the *R* property is being red S-*simpliciter*, then we should judge that the ball is manifestly red at t_3. So being red M-*simpliciter* and having the second-order conjoined properties do not provide the truth conditions for being red S-*simpliciter*, or for our utterances of 'the ball is manifestly red.'

This problem prevails regardless. For suppose that the *R* property is the property of having the property of being red $t_1$ly at t_1, and red $t_2$ly at t_2. Call this property the *R** property. This seems more promising; after all, the ball is red at t_1 and t_2 in virtue of being red $t_1$ly at t_1 and red $t_2$ly at t_2. But what is this property of being red $t_1$ly at t_1? Does the ball have the property of being red $t_1$ly at t_1 at times other than t_1? Does the ball have the property of being red $t_1$ly at t_1, *at* t_3? The very same arguments that we earlier rehearsed which tell us that the ball has the property of being red $t_1$ly at every time at which it exists, will also tell us that it must have the property of being red $t_1$ly at t_1 at all times at which it exists. Hence it must indeed have the property of being red $t_1$ly at t_1 at t_3. But then we are back to where we began, with the ball instantiating the *R** property at t_3 when we want to say that it is not red S-*simpliciter*.

4.1.1 Variable Role Adverbialism

What sense can we make of the notion of having a property S-*simpliciter*? Well what explains why the ball is not manifestly red at t_3 is that while the ball possesses redness $t_1$ly, $t_2$ly and $t_4$ly, it fails to possess it $t_3$ly. The lesson is that for the adverbialist, the semantics of property talk can be understood as analogous to the semantics of first-order functionalism in the philosophy of mind. Recall that for the adverbialist, the ball is red M-*simpliciter*, so it possesses the property of redness at every time at which it exists, including those times at which we would not describe the ball as red, namely at t_3.

It is not sufficient for something to be red S-*simpliciter* that it possess the property of redness. Being red M-*simpliciter* is a necessary, but not a sufficient condition, for something to be manifestly red at a particular time. The ball is manifestly red only when the property of redness plays a particular functional role—the role of causing the ball to appear red—what we might call the manifest-redness role. Or, to use a more perspicuous example, suppose that at t_4 the ball is flattened. Then suppose that at t_3 the ball has the property of being round—it is round M-*simpliciter*. At t_4 we do not judge that the ball is round. Why not? Because at t_4 roundness is not playing the appropriate functional role, namely the role of causing things to roll, to be circular, and so forth. While roundness is a necessary condition for something to count as being round S-*simpliciter* at some time, it is not sufficient. It is also necessary that at that time, the roundness property play the appropriate role—namely the manifest-roundness role.

What it takes for it to be correct to judge at some time t that an object O has property P—has P S-*simpliciter*—is for O to have P M-*simpliciter*, and for the P property to play the appropriate role at t. What is it for P to play the appropriate role at t? Return to the ball. Why does redness play the manifest-redness role at t_1, t_2 and t_4, but not t_3? Well talking of a univocal 'manifest-redness role' is a little misleading. For there is no single role that redness plays, in virtue of which redness is made manifest. We should talk about the functional *roles* that redness plays. What are these roles? They are the roles of being instantiated in particular temporal ways. At t_1 the role of being manifestly red is the role of being instantiated $t_1$ly. At t_1 the ball is red S-*simpliciter* because redness plays a particular functional role: the role of being instantiated $t_1$ly. At t_2 redness must play a different role in order to be manifest: the role of being instantiated $t_2$ly. Redness is not manifest at t_3 because at t_3 the appropriate functional role is not played: at t_3 redness does not play the role of being instantiated $t_3$ly. In general, what it is for some property to play the appropriate role at some time t_n, is for that property to have the second-order property of being instantiated $t_n$ly. It is having these second-order properties that explains an object's appearing, or failing to appear certain ways at different times: for it is in virtue of these second-order properties that first-order properties are made manifest at times.

This account then, is analogous to a first-order functionalist account of mental properties, but is instead a first-order functionalist account of the instantiation of properties S-*simpliciter*. To clarify, consider the property of being in pain. First-order analytic functionalists think that pain is just whatever plays the pain role, and thus there may be nothing in common between the various realisers of pain aside from the fact that they are pain realisers.[25] First-order empirical functionalists think that pain is whatever actually plays the pain role—they think 'pain' is a rigid designator—and thus if C fibres firing actually play the pain role, then C fibres firing are all and only the pains.[26] So there is something in common between all of the realisers of pain, namely that they are all C fibres. Variable role adverbialism falls somewhere between these two views. Just as empirical functionalists are wrong to think that being a C fibre firing is sufficient for something to count as being a pain—no C fibre firing in a scientist's petrie dish is a pain—so too on this view having the redness property at a time is not sufficient for us correctly to judge that something is manifestly red at that time. But, unlike the analytic functionalist who holds that any realiser may realise pain, on this view only the property of redness can play the appropriate functional role such that redness is made manifest at that time. So the existence of the redness property is *necessary* for any object to count as being red S-*simpliciter*.

This view is analogous to a view that one might have about pain: that pain is C fibres firing just when those fibres play the pain role. Thus only C fibres are ever pains, but sometimes C fibres do not realise pain, namely when they are in petrie dishes and are not playing the pain role. This is a first order functionalism because where the C fibres are playing the pain role it is the fibres *themselves* that are the pain, not the second-order property of being a property that plays the pain role, even though it is in virtue of that second-order property's instantiation that the fibres are the pain.

Variable role adverbialism is also a first-order functionalist account despite the fact that it appeals to second-order properties as the role determining properties. On this view the ball is red S-*simpliciter* at a time just if at that time it has the first-order property of redness and that first-order property plays the relevant functional role of making redness manifest. The fact that what it *takes* for redness to play the relevant role is for that first-order property to have the property of being instantiated in a particular temporal manner does not turn this into a second-order functionalist account.[27]

Returning to the ball then, whenever the ball is red S-*simpliciter*, it is in virtue of the very same property being instantiated—redness—and that property playing the appropriate role at that time. So at any time when the ball is red S-*simpliciter* the ball is also red M-*simpliciter*—but the property of being red M-*simpliciter* only counts as being red S-*simpliciter* where that property also plays the appropriate role. Being red S-*simpliciter* tracks the same property across time, it is just that it only tracks that property when it plays the relevant role.

Thus the ball is red S-*simpliciter* at some time *t* iff:

(a) the ball has the property of being red M-*simpliciter*.

and

(b) at *t* redness plays the relevant functional role.

Compare this to the truth conditions for our earlier example of pain. In that case *X* is in pain at some time *t* iff:

(a') *X* has the property of having C fibres firing.

and

(b') at *t* the C fibres play the relevant functional role.

In both these cases, pain and redness are manifest at a time just if the relevant property is playing the relevant role at that time. But there is a crucial difference between the cases, and that difference emerges when we analyse conditions (b) and (b'). Consider (b'): at *t* the C fibres play the relevant functional role. What is that role? The pain role. Suppose, for simplicity, that what it is to play the pain role is to be caused by bodily damage and to seek to avoid such damage. Then on every occasion in which *X* is in pain, it is in virtue of C fibres playing one and the same role: the pain role.

That is not the case when we consider the variable role adverbialist account of properties. Consider condition (b): at *t* redness plays the relevant functional role. What is the relevant functional role? Well, which role redness must play in order for it to be made manifest at a time is sensitive to temporal facts. At each distinct temporal location it is a different role that redness must play in order to be manifest at that time: at t_1 it is the role of being instantiated $t_1$ly, and at t_2 it is the role of being instantiated $t_2$ly and so forth.

If we return to the pain case we can construct an example—albeit fictional—that is analogous to the one we find in the adverbialist case. Suppose that what it is to play the pain role is to have a certain mass, say 3 nanograms. Then all and only the C fibres that have a mass of 3 nanograms are pains. There is just one functional role—having a mass of 3 nanograms—that is the pain role. Now suppose we alter the example slightly so that the mass that is required for something to count as being pain varies depending on where in the brain the C fibre is located. What it is to be pain is to have a mass of 1 at location 1, a mass of 2 at location 2 and so forth. So a C fibre at location 5 will count as being a pain only if that C fibre has a mass of 5 nanograms. So at each different location, a different role must be played for a firing C fibre to count as a pain. At location 1 it is the role of having a mass of 1 nanogram, at location 2 it is the role of having a mass of 2 nanograms and so on: there is no univocal pain role.

There is, however, an important meta-role at play here. For there is surely something important in common between each of the roles just described. What it is to be a pain is to be in location N and have a mass of N nanograms. And to grasp what it is to be pain is not to grasp any of the roles that have to be played for something to count as a pain at *particular* locations—it is to grasp something more abstract, the meta-role. It is to grasp how the role that must be played depends on the location. The reason that at different locations it is different roles that must be played, is because the meta-role—having C fibres with a mass of N nanograms at location N—has a hidden spatial indexical. And thus playing the meta-role involves playing different roles at different locations.

Something analogous is true on the variable adverbial account of properties. In that case, although it is different roles that must be played at different times for properties to be manifest at those times, there is a meta-role that is in common between every instantiation of a property S-*simpliciter*. What it is for redness to be manifest at a time t_n, is for redness to be instantiated $t_n$ly at t_n. So the meta-role here is that a property P is instantiated S-*simpliciter* at t_n just if P is instantiated $t_n$ly at t_n. Instead of a spatial indexical we find a temporal indexical such that playing the meta-role involves playing different roles at different times. And once again, it is grasping this meta-role role that is important to grasping the idea of having a property S-*simpliciter*. What matters in grasping the idea of being red S-*simpliciter* is not that one grasps that redness must be instantiated $t_1$ly at t_1[28] to be made manifest, nor that it must be instantiated $t_2$ly at t_2 to be made manifest, but rather, what must be grasped is that the manner in which redness must be instantiated is sensitive to which time it is: namely that redness must be instantiated in the same temporal manner as the current temporal location.

4.1.2 A-intensions and the Semantics of S-simpliciter

So far we have explicated the metaphysical component of the notion of instantiating a property S-*simpliciter*. What are we to say though, of what is in common between our judgements at different times that a property is instantiated S-*simpliciter*? The idea is that what is in common between all our judgements that the ball is red S-*simpliciter*, is the relevant meta-role. One way to further explicate this notion of common content between judgements about properties instantiated S-*simpliciter*, is by comparison with two-dimensional semantics.

Briefly, the core idea of two-dimensional semantics is that sentences have different intensions when considered along two different semantic dimensions.[29] On one dimension, what Jackson calls the *C*-intension,[30] we consider what terms pick out in worlds considered as counterfactual. The *C*-intension is what we might think of as being the 'Kripke intension.' If water in the actual world is H_20, then considered counterfactually 'water' picks out all and only H_20 in all other worlds. The other intension, what Jackson call the *A*-intension, is the dimension along which we consider what terms pick out in worlds considered as actual. If the actual world is one in which a clear potable liquid of somewhat different chemical composition than H_20 exists, then 'water' refers to that substance. So if considered as actual, the chemical composition of that liquid is XYZ, then 'water is XYZ' is true. The *A*-intension tracks what is semantically in common between utterances of 'this is water' in different worlds considered as actual.

Earlier we introduced the idea of a meta-role, and that this is crucial in grasping the idea of instantiating a property S-*simpliciter*. So what is in common between judgements at different times that a property is instantiated S-*simpliciter*, is that the same meta-role is being played at each of these times. For instance, we could say that the truth conditions for an utterance of 'the ball is red S-*simpliciter*' made at any time t_n are as follows:

(a) the ball has the property of redness M-*simpliciter*

and

(b) the property of redness is instantiated $t_n$ly at t_n.

Given that (b) has a temporal indexical, however, in what sense is there any content in common between an utterance of 'the ball is manifestly red' made at t_1, and an utterance made at t_2. At t_1 it is because redness plays the role of being instantiated $t_1$ly that the ball is manifestly red, and at t_2 it is in virtue of redness playing a different role, the $t_2$ly role, that the ball is manifestly red. But just as 'the ball is manifestly red' is true at different times in virtue of different roles being played at those times, so too the *A*-intension of 'this is water' picks out different chemical substances in different worlds considered as actual. This suggests that we might employ a temporal analog of the *A*-intension to explain what is semantically in common between utterances that attribute properties S-*simpliciter*. In that case we would evaluate statements such as 'the ball is red' at different times considered as now, rather than different worlds considered as actual. Let us call the temporal analog of the A-intension the *temporal A-intension.*

Consider the case of water: what underlies the *A*-intension of the claim made of some liquid *L*, that it is water? Let us say that *L* is water just if *L* is a sample of something that has properties that play the water-role—properties such as being clear, potable, liquid and so forth.

Under what circumstances is the ball now red S-*simpliciter*? If it now has properties that make it manifest that the ball is red: that is, if it now has the property of redness and that property now plays the appropriate functional role—what we might call the t_{now}ly role. Hence for every time considered as now, we rightly judge that the ball is red just if redness plays the t_{now}ly role: that is, if redness is instantiated $t_n$ly at t_n. Thus at t_1, 'the ball is (manifestly) red' is true just if redness plays the $t_1$ly role, and at t_2 is true just if redness plays the $t_2$ly role. It is this that explains how utterances such as 'the ball is (manifestly) red' can be true at one time (t_1 and t_2) and false at another time (t_3) despite the fact that the ball is strictly identical across time and thus has all of the same properties at each of those times. For the temporal *A*-intension picks out different propositions at different times: at t_1 it picks out the proposition 'instantiates redness $t_1$ly' and at t_2 picks out a different proposition 'instantiates redness $t_2$ly', and there is nothing contradictory in this.

In effect then, the temporal *A*-intension is the meta-role of having redness play the relevant functional role. This means that there is an important difference between *A*-intensions as they are usually conceived, and temporal *A*-intensions. In the former case we will say, for instance, that what it *is* that plays the water role (H_20, XYZ, etc) varies depending on which world we take to be actual. But the water role, the role of being clear, potable, liquid and so forth, remains constant across worlds considered as actual. In the case of temporal *A*-intensions, it is not that what plays the role of rendering redness manifest varies depending on which time we take to be now: for it is always redness that plays that role. Rather, what varies across time is the role itself.

To clarify this, we can imagine a case in which a similar phenomenon presents itself when considering traditional *A*-intensions. We have been supposing that the *A*-intension of 'water' is something like, 'water is whatever actually plays the water role'. But suppose our semantic intuitions were radically different. Suppose instead we held that the *A*-intension of water contained 'world indexicals', such that the role that some substance has to play in order to count as being water depends on which world the substance is in. So for instance, if we are in world w_1, then water is whatever plays the R_1 role. If we are in w_2, then water is whatever plays the R_2 role. As we move from world to world considering each as actual, it is different roles that we must consider when determining the referent of 'water' in each world. Nevertheless, there is still something in common between any utterance of 'this is water' made at any world considered as actual: namely that something is water only if plays role R_n at world w_n. This role remains constant across worlds, and it captures the A-intension of 'water' in this case.

In the same way, it is the role of instantiating properties $t_n$ly at t_n that remains constant across times, and it is this that is the content in common between judgements at different times, that something possesses a property S-*simpliciter*. Of course, in this case the temporal A-intension is much more illuminating than the crazy one we just considered. In this latter case the link between which world one is in, and which role some substance ought play to count as water, appears tenuous to say the least. So we might be tempted to say that here, the A-intension of 'water' captures only a very thin common content. Not so, however, in the more plausible temporal case. For we might expect that what time it is will be intimately linked to the manner in which a property needs to be instantiated in order to be manifest at that time. Moreover, it is grasping the meta-role of instantiating properties $t_n$ly at t_n that is at the heart of an understanding of what it is to instantiate a property S-*simpliciter*: for what is crucial is to understand that there is a particular relation that needs to hold between the time and the manner of instantiation of a property if that property is to be manifest at that time. That utterances of 'the ball is red' have semantic content prior to the *a posteriori* discovery of which time is now, and thus which role redness must play in order to be manifest, attests to the importance of the temporal A-intension.[31]

4.2 Having Parts *Simpliciter*

So how does variable role adverbialism help when it comes to the issue of persisting objects having different *parts* at different times? The problem, is that we want to say that in some sense three-dimensional objects have all of their parts present whenever they exist. Yet we cannot, it seems, say that three-dimensional objects have all of their parts *simpliciter* present whenever they exist, since persisting objects have different parts at different times. Perhaps though, an analogous distinction to the one we just made in the case of property instantiation can be drawn with respect to parts: perhaps we can distinguish between *having a part* M-*simpliciter* and *having a part* S-*simpliciter*. And perhaps that distinction can help us in defining what it is to endure.

In what follows I will argue that in an abstract way we can draw a distinction between two ways of having a part *simpliciter*, and we can then use the notion of having a part S-*simpliciter* in defining endurance. However, the exact details of these two different notions of *simpliciter* when applied to parthood will depend on which other views about composition the three-dimensionalist embraces. These additional views will be discussed in chapter four, and then in chapters five and six I will return to consider fully the way three-dimensional objects have parts M-*simpliciter* and/or S-*simpliciter*. For now I will merely sketch the broad idea behind these two conceptions of having a part *simpliciter*, so that we can employ the distinction in our definition of endurance.

It is sometimes said that for the three-dimensionalist, parthood is never atemporal: we can only talk of enduring objects having parts at times, we cannot talk about the parts of enduring objects *simpliciter*. This view is a sort of analog of indexicalism, according to which enduring objects have what we might think of as parts-at-times. To illustrate, suppose that there exists some object O that is at t_1 composed of proper parts A and B, and at t_2 composed of parts A and C. Then the indexicalist will say that O has the property of having-B-as-a-part-at-t_1. There is no sense in which O just has the property of having B as a part. Similarly, there is no real sense in which O just *has* B as a part: O only has B as a part at a time. O does not have B *simpliciter*.

The adverbialist, however, will hold that O does straightforwardly have the property of having B as a part, and has that property *simpliciter*, it is merely that O has that property in a $t_1$ly manner. Thus O has the property of having B as a part M-*simpliciter*: it has that property at every time at which it exists. So too, we might say that for the adverbialist, rather than having parts-at-times, instead persisting objects have parts, but they have them in different temporally modified ways. O straightforwardly has part B, it is just that O has B in the $t_1$ly way. Persisting objects do not have parts-at-times (having a part is not a disguised relation to a time), rather, such objects just have parts, they have them in different temporally modified ways.

Then just as a commitment to underlying properties that are instantiated *simpliciter* in different ways at different times commits one to holding that these underlying properties are possessed at all times though they may only be manifest at some of those times, so too a commitment to parts that are had *simpliciter* in different ways at different times will commit one to holding that these parts are possessed at all times, though they may only be manifest at some of those times. Call the sense in which if an enduring object O has some part *simpliciter*, then it has that part at all times at which it exists, *having a part M-simpliciter.*

Then just as our ball instantiates redness M-*simpliciter* at t_3 despite being manifestly blue at that time, so too O has part B at t_2 despite not manifesting the part at t_2. What it is to manifest a part is to have a part S-*simpliciter*. And just as our usual property attributions track the instantiation of properties S-*simpliciter*, so too our usual attribution of parts tracks the having of those parts S-*simpliciter*. So P is part of O S-*simpliciter* at t just if at t an ideal observer would judge, at t, that P is part of O. And just as variable role adverbialism tells us that a property P is instantiated S-*simpliciter* at some time t just if at t, P plays the appropriate role, the same will be true for having parts S-*simpliciter*. Namely, O has some part P S-*simpliciter* at t just if O has P M-*simpliciter* and P plays the appropriate role at t. The role that P must play will, of course, vary depending on the time. At t_1 P will need to play the $t_1$ly role, and *mutatis mutandis* for other times. Thus at different times, P plays the appropriate role at that time in virtue of having the property of being had in a particular temporal manner. In fact, we could say that what it is for O to have P S-*simpliciter* at t, is for P to have the property, at t, of being had $t$ly by O (that is, P has the property of being a $t$ly part of O) or we could say that what it is for O to have P S-*simpliciter* at t is for O to have the property at t of having P $t$ly (that is, O has the property of having part P $t$ly).

Since it is having a part S-*simpliciter* that tracks our judgements about parthood, this explains how utterances such as 'B is part of O' can be true at one time (t_1) and false at another (t_2) even though O is strictly identical across time. At t_1 the claim is made true by B playing the $t_1$ly role, and at t_2 is made false by B failing to play the $t_2$ly role. But the properties that O has at each time remain the same, namely the properties of instantiating the B property M-*simpliciter* and of instantiating the B property $t_1$ly. Furthermore, as in the case of variable role adverbialism, there is some meta-role in common between judgements made at different times, that, say, 'B is part of O'. Consider A. A is part of O S-*simpliciter* at t_1 and t_2. At t_1 it is manifestly part of O in virtue of playing the $t_1$ly role, and at t_2 in virtue of playing a different role, the $t_2$ly role. But there is a meta-role in common here, the meta-role of playing the role of being had in the same manner as the current time: the $t_n$ly role at t_n. This is the role that is in common between any two times at which some part is had S-*simpliciter*. So some P is part of O S-*simpliciter* at any time t_n, just in case P is part of O M-*simpliciter*, and at t_n P is part of O $t_n$ly.

What can it really mean though, to talk of the existence of parts that are not manifest at certain times? In part this is a general worry faced by any three-dimensionalist account that holds that there is any sense in which properties or parts are had *simpliciter*, but in different ways at different times. For then one must hold that instantiating a particular property, or having a part *simpliciter*, is not sufficient for some enduring object to manifest that property or that part at some time. As we will see in chapter five, however, the manner in which the notion of a manifest or non-manifest part is cashed out will depend on other details of the three-dimensionalist view in question. So I will put aside this issue for the moment and turn instead to develop a schematic definition of endurance: schematic in the sense that it makes reference to having parts S-*simpliciter*, where what it is to have a part S-*simpliciter* will vary slightly depending on the particular version of three-dimensionalism in question. Thus we have the following:

> END: An object *endures* iff it exists at multiple times, and is wholly present at each of those times, where an object is 'wholly present' at a time just in case all of its parts S-*simpliciter* are present at that time.

This definition of endurance captures the core three-dimensionalist intuition that in some sense all of an enduring object's parts are present whenever it exists. Consider object *O* once more. At t_1 *O* is composed of proper parts *A* and *B*, and at t_2 is composed of parts *A* and *C*. So at t_1 *A* and *B* are all and only the parts that *O* has S-*simpliciter*: at t_1, *O* does not have part *C* S-*simpliciter*. So O is wholly present at t_1 in virtue of having all of its parts (*A* and *B*), S-*simpliciter* at t_1. Furthermore, there is a clear sense in which this definition sheds light on the issue of persistence. Unlike the possibilist definitions we considered earlier, it tells us something about how objects persist in the actual world, not just how they might persist in some possible world. It does not just tell us that actual objects (if actual objects endure) possibly have no temporal parts, rather, it tells us that they do not have temporal parts.

How does this definition distinguish endurance from perdurance and terdurance? *Prima facie* at least, the four-dimensionalist will reject the idea that properties or parts are had M- or S-*simpliciter*. For these notions are defined in terms of temporal adverbialisation, and this is a strategy that four-dimensionalists explicitly reject. So no four-dimensionalist is going to agree that four-dimensional objects have all of their parts S-*simpliciter* at times, for four-dimensional objects do not have parts S-*simpliciter*.

Notice also that for the three-dimensionalist, having a part S-*simpliciter* is supposed to capture the everyday sense of having a part, the straightforward sense in which we judge that *P* is part of *O*. Put aside for a moment the specifics of the three-dimensionalist account of what it is to have a part S-*simpliciter*, and instead think of having a part S-*simpliciter* as just having a part in this everyday straightforward sense. Now, the perdurantist thinks it is tenselessly true that *AB* (the fusion of *A* and *B*) is a temporal part of *O*, and that *AC* (the fusion of *A* and *C*) is a temporal part of *O*. She also thinks it is tenselessly true that *A*, *B*, and *C* are (temporal) parts of *O* (albeit non-maximal temporal parts). So she thinks that at t_1 not only is the utterance '*A* is part of *O*' true, but so too is the utterance '*C* is part of *O*'. So if what it is to have a part S-*simpliciter* is to have a part in the everyday straightforward sense, then the perdurantist would say that at t_1, *C* is part of *O* S-*simpliciter*. Plausibly, the terdurantist will agree. Although the terdurantist denies the existence of *AB* and *AC*, she clearly does not deny that terduring objects have spatial parts at times. Although many of the details of terdurantism remain to be explicated, it is plausible that the terdurantist will hold that terduring objects have their parts tenselessly: after all terduring objects will likely turn out to be fusions of terduring simples. In that case, the terdurantist too will hold that if *O* is a terduring object, then at t_1 *C* is part of *O*. So the terdurantist will agree with the perdurantist that *C* is part of *O* S-*simpliciter* at t_1. For both perdurantist and terdurantist hold that four-dimensional objects are composed of non-present parts: that is what it is to be a four-dimensional object that is 'spread out' in time. The three-dimensionalist disagrees; she holds that *C* is not part of *O* S-*simpliciter* at t_1. She thinks that in the everyday sense of 'having a part', it is false that *C* is part of *O* at t_1.

So we can see that both perdurantist and terdurantist would hold that four-dimensional objects have parts S-*simpliciter* at times other than at the times at which those parts are manifest: for both perdurantist and terdurantist hold that four-dimensional objects are composed of non-present parts. The three-dimensionalist disagrees, and thus defines the notion of having a part S-*simpliciter* accordingly, and it is this that distinguishes endurance from either perdurance of terdurance.

Finally, is our definition of endurance consistent with the ancillary metaphysical commitments that we met earlier? Consider first presentism and eternalism. The definition is essentially constructed with an eternalist view of time in mind: after all, it borrows from the adverbialist account of property instantiation, which is an account designed to reconcile eternalism, persistence and strict identity. It is invoking the distinction between having a part M- and S-*simpliciter* that allows the three-dimensionalist to hold that there is a sense in which enduring objects have all of their parts *simpliciter* present whenever they exist. So the definition is certainly consistent with eternalism. It is also compatible with presentism. If presentism is true, then the only time that is ontologically real is the present. So whatever exists, must exist in the present. Hence at whatever time an enduring object *O* exists, it is trivially true that all of its parts S-*simpliciter* exist at that time.

This definition can also accommodate a range of views about composition. This is not surprising, since the definition tells us nothing about which composite objects exist. It does not, for instance, rule out the possibility that there exist temporally scattered objects that endure, nor does it rule out the possibility that there exist odd gerrymandered objects such as trout-turkeys, (a fusion of a trout and a turkey) which, if they persist, do so by enduring. So the definition is compatible with either restricted or unrestricted composition. All it tells us is that if this is a world where objects endure, then it is a world in which no persisting objects are composed of non-present parts S-*simpliciter*.

Now that we have the key terms of three- and four-dimensionalism defined, we can move on in chapters five and six to argue that there is a correct translation between 'analogous versions' of each theory. But what are these 'analogous versions' of three- and four-dimensionalism? Well, we know that in general they are versions of each theory that adopt the same set of ancillary metaphysical commitments. In large part, they are versions of each theory that adopt different views about composition, and different views about the nature of simples. In the next two chapters I develop a number of different versions of three- and four-dimensionalism. What is revealing about these theories is what they tell us about the relationship between persistence and composition, and what sorts of problems they reveal for certain classes of view that embrace a particular package of metaphysical views. Moreover, what is interesting is that many of these theories are entirely novel accounts of persistence that provide quite different resources for dealing with the puzzle cases, than do the traditional versions of three- and four-dimensionalism that we have already met.

NOTES

[1] In chapter four I explore fully the relation between this rough understanding of restricted and unrestricted composition, and the more explicit formulation of mereological universalism.
[2] Sider (2001) p 60.
[3] Sider (2001) p 60.
[4] Armstrong (1980).
[5] Or at least, it is analogous to a particular view one might have about such spatially extended simples. One could treat such simples in a way analogous to the way that the three-dimensionalist treats persistence. One could say that extended simples wholly exist at each location at which they exist. Then such simples are literally strictly identical across space, the way that enduring objects are strictly identical across time. I consider a view a little like this in chapter seven. I don't think that this is the usual view about such simples (though it is not usually stated which of these two views defenders of the idea of such simples actually adopt), but it does not matter: the conception I outline above is clearly coherent.
[6] Brogard (2000).
[7] One might wonder then, what the difference is between enduring and perduring. After all, (instantaneous) temporal parts are wholly present whenever they exist: so if all that ever exists

is a temporal part, in what sense is the object in question four, rather than three-dimensional? To be sure, I do not think that this is an intuitive view. But I suppose the idea is that each of these temporal parts is distinct. At each time at which a persisting object exists, we see a distinct object. In fact, we can never see the whole object, since only a part of it ever exists. But this is unlike the three-dimensionalist who thinks that each three-dimensional object that exists at each successively present moment, is the very same strictly identical object.

[8] For further discussion of the stage view see Sider (2000). For arguments in favour of the worm view see Balashov (2002).

[9] Sider (2001). chapter three.

[10] Merricks (1999).

[11] See for instance Wiggins (1968); Baker (1997); Johnston (1992).

[12] Merricks (1995).

[13] Merricks (1999).

[14] Markosian (1994).

[15] See for instance Thomson (1983).

[16] Van Inwagen (2002) writes: "Since I understand all the words, I understand 'Lewis-part" and know what Lewis-parts are. In a way. In the same way as the way in which I should understand talk of "propertyless objects" if I were told that "propertyless object" meant "object of which nothing is true." ...But I should hardly care to say that I know what someone was talking about...who talked of them in a way that suggested that he supposed there were such things." (Where a Lewis-part is a temporal stage of Lewis). p 445.

[17] Sider (2001) p 65.

[18] Sider (2001) p 66.

[19] Lewis (1986) p 204.

[20] Stone raises a similar problem in his (2003). Of course, the indexicalist can go some way towards explaining what is in common between different instantiations of redness-at-t, by noting that in each case, we have the same relation—redness-at. We have the same relation, but a different property. Still, this doesn't really explain why in cases where we would traditionally have said that we have an intrinsic property, (as in the case of redness) the temporal relata seems to make no difference to how things appear. Consider the case of a normal relation, say, next to. The property next-to-Bill looks very different to the property next-to-Fred, even though in both cases we have the same relation. There is a commonality (the next-to commonality), but the properties look different. Yet the various redness-at-t properties look the same regardless of changes in the relata.

[21] Johnston (1987).

[22] Johnston certainly sees this, others concentrate on persisting traits while they are manifest, ie. redness at t_1 and t_2.

[23] Johnston uses a similar analogy in his (1987).

[24] The idea of second-order properties is formulated here as properties particulars may possess in virtue of having certain properties or having them instantiated in certain ways. But the argument could equally be formulated in terms of the other conception found in the literature, as a property of a property. In this latter case we would think of being instantiated $t_2$ly as a property of the enduring property instance of redness, which it must still possess at t_3 on pain of contradiction.

[25] Lewis (1972); Braddon-Mitchell (2003).

[26] Loar (1981); Lycan (1990).

[27] Braddon-Mitchell and Jackson (1996).

[28] This formulation works because redness is instantiated $t_1$ly at all times if instantiated at all: the more cumbersome at $t_1$redness would need to be instantiated $t_1$ly at t_1 would be required if this were not so. What changes is not whether redness is instantiated $t_1$ly; it is what kind of instantiation is relevant.

[29] Frank Jackson uses the terminology of an A-intension, while David Chalmers uses the terminology of a primary intension. For more on two-dimensional semantics see Jackson (2004); Braddon-Mitchell (2004) and Chalmers (2004).

[30] Jackson (2004).

[31] Are there any temporal C-intensions? If there were, they might be formed by temporal rigidification. If it is now t_1, we know that redness would need to be had $t_1$ly for a ball to be red S-*simpliciter*. Rigidifying on that, we would say that for a ball to be red S-*simpliciter* at other times regarded as countertemporal, redness would need to be possessed $t_1$ly. But redness always is possessed $t_1$ly, so the ball would be always red S-*simpliciter* if it is ever red S-*simpliciter*.

The moral of this is that we should not temporally rigidify. It gives the wrong answers. But there is an interesting reason why this is so: it is because we expect to find other times to be now, whereas we do not expect other worlds to become actual that are not! If there were world travel, we would not rigidify (or at least the pressure to do so would be reduced). If next year we expected to be in a world where XYZ played the water role, then we would be more inclined to just say that water is whatever plays the water role, rather than that water is whatever actually plays the water role: in the former case the *A* and *C* intensions do not come apart.

But there is 'time travel', we do exist at multiple temporal locations. We expect to be in the future, thus there is good reason to suppose that temporal *A* and *C* intensions will not come apart. So we are inclined to judge whether the ball is red S-*simpliciter* by the standards that are appropriate for that time considered as now, as that is when the judgement will be crucial.

Chapter 4

ISSUES OF COMPOSITION

1. ANCILLARY COMMITMENTS CONSIDERED

The view that composition is restricted is the view that composition only occurs under certain circumstances. The view that composition is unrestricted is the view that composition occurs under any circumstance. There are various versions of restricted composition that restrict composition in such a way that only *some* of the everyday objects that we recognise are held to exist (for instance the view that composition only occurs when simples are arranged in such a way that they compose something that has a life).[1] The most natural version of restricted composition, however, is one that rejects the idea that there exist odd gerrymandered objects such as the object composed of my shoe and your dog, while accepting that all of the ordinary objects of our experience exist. The view that composition is unrestricted, on the other hand, is the view that there exists a plethora of peculiar objects many of which have as parts the sorts of objects with which we are familiar.

In chapter three I claimed that both three- and four-dimensionalism are consistent with either restricted or unrestricted composition. That claim, however, remains to be defended. For of late it has been argued that the view that composition is unrestricted inevitably leads to the view that objects are four-dimensional. If true, that would put at risk the thesis that the two theories are equivalent.

In section 2 of this chapter I begin by outlining an influential recent argument developed by Ted Sider. Sider's argument purports to show that objects are four-dimensional and perdure, via the claim that composition is never vague and thus that composition is unrestricted. I argue that even if we grant that vagueness is only and always the result of semantic indeterminacy and thus also grant that composition is unrestricted, it does not follow that objects persist by perduring. This is true for two reasons. The first, which I will explore in section 3, is that the thesis that composition is unrestricted does not entail that there exist instantaneous objects that wholly overlap persisting objects at times, and thus does not entail that there exists anything that could be called a temporal part. In this section I introduce two analogous versions of three- and four-dimensionalism—what I call *unitary three-dimensionalism* and *unitary four-dimensionalism*, each of which reject the idea that there exist such instantaneous objects.

The second reason why embracing unrestricted composition does not lead inevitably to accepting four-dimensionalism, is that even if we grant that there exist all of the various instantaneous objects that overlap persisting objects at times, it does not follow that such persisting objects perdure. To show this, in section 4 I outline a coherent version of three-dimensionalism that grants just such an assumption—a view I call *unrestricted non-unitary three-dimensionalism.* This view is analogous to an unrestricted version of four-dimensionalism: *unrestricted non-unitary four-dimensionalism.*

In part then, developing these different analogous versions of three and four-dimensionalism is part of the project of showing that these two theories are metaphysically equivalent. For when I argue, in the following chapters, that three- and four-dimensionalism are equivalent, I am arguing that versions of each that adopt the same set of ancillary commitments are equivalent. That is, I am arguing that *analogous* versions of each are equivalent. The hope, however, is that in developing these theories we will discover new insight into the ways in which one's views about persistence, composition, and the nature of the fundamental furniture of the universe, are deeply interrelated.

So let us proceed to outline and consider the argument from vagueness, which aims to show that three-dimensionalism is inconsistent with unrestricted composition. Though I contend that this argument fails, seeing what is right about the argument will provide the impetus to develop new versions of three-dimensionalism that embrace alternative views about composition.

2. VAGUENESS AND UNRESTRICTED COMPOSITION

In his recent book, Ted Sider[2] contends that the best argument in favour of four-dimensionalism is the argument from vagueness. This argument has two parts: the first part seeks to establish that composition is unrestricted, via the claim that composition is never vague. The second part seeks to show that if composition is unrestricted, then it follows that persisting objects are composed of temporal parts, and thus that perdurantism, (and hence four-dimensionalism) is true. I argue that even if we accept the first part of the argument, we are not led inevitably to accept four-dimensionalism. Rather, considerations pertaining to vagueness are strictly orthogonal to the issue of the manner in which objects persist.

2.1 The Argument from Vagueness

The first part of the argument from vagueness owes its origins to David Lewis.[3] Lewis argues that any attempt to restrict composition in a way that is in-keeping with our intuitions about which objects exist, must necessarily be a vague restriction. For commonplace objects all have imprecise temporal and spatial borders, and thus there is no determinate point at which, for instance, some atom *A* can be said to be part of some object *O*, or not part of *O*. But if composition itself is vague, then existence will be vague. It will be vague at exactly which moment an object comes into existence, and vague at exactly which moment it ceases to exist: for every object, there will be some time at which it is indeterminate whether that object exists or not. Since Lewis thinks that vagueness is never ontological,[4] but rather is the result of semantic indeterminacy,[5] he holds that existence cannot be a matter of degree, and thus he concludes that composition cannot be restricted. Recently, Lewis' argument for unrestricted composition has been further refined by Sider.[6] Expressed as a reductio, Sider's argument is as follows.

Part I: From Vagueness to Unrestricted Composition

1. Assumption: Existence is not vague: composition either definitely occurs or definitely does not occur.
2. Assumption: Not every arrangement of matter composes an object.
3. So there must be a continuum of cases such that at one end of the continuum composition occurs, and at the other end composition fails to occur.
4. Each of the cases on the continuum is highly similar to the adjacent cases.
5. So there is no principled way to draw the line between a case where composition occurs and a case where composition does not occur.
6. So there are cases where it is indeterminate whether composition occurs or not.
7. So existence is vague.

As it stands, this argument requires the rejection of either (1) or (2). In general, arguments of this form have been resisted by disputing the truth of (1).[7] Hence Sider has constructed a number of ancillary arguments that aim to bolster (1) and thus lead to the rejection of (2).[8] In this chapter, however, I want to focus on the second part of the argument from vagueness, namely that part that moves from the falsity of (2) to the truth of four-dimensionalism. Let us turn to this second part of the argument from vagueness. Here is a reconstruction of Sider's argument:[9]

Part II From Unrestricted Composition to Temporal Parts

1. Assumption: Composition is unrestricted.
2. So there is a fusion of the members of any arbitrary set S at time t, where x is a fusion of the members of S at t iff every member of S is part of x at t, and each part of x at t overlaps at t some member of S.[10]
3. Objects persist through time, so we need a temporalised version of unrestricted composition.
4. An object x is a diachronic fusion of the members of sets S_1, S_2, and S_3 at times t_1, t_2 and t_3 if (i) x is composed of S_1 at t_1, S_2 at t_2, and S_3 at t_3 and (ii) x exists only at the times t_1 t_2 and t_3.[11]
5. Since composition is unrestricted, any sets of objects at times has a diachronic fusion. Following Sider, call this thesis U: Any arbitrary sets S_is and times t_is has a diachronic fusion.[12]
6. Given U, it follows that there for any y at t, there is some instantaneous object x that (i) is the fusion of the members of set y at t, and (ii) which exists only at t.
7. y is part of x at t and every part of x at t overlaps y at t (from (i) and the definition of fusion).[13]
8. With certain mereological principles we can then move from the claim that every part of x at t overlaps y at t, to the claim that x is part of y at t.[14]
9. Now consider the definition of an instantaneous temporal part: x is an instantaneous temporal part of y at instant $t =_{df}$ (a) x is part of y at t (b) x exists at, but only at t and (c) x overlaps at t everything that is part of y at t.[15]
10. The diachronic fusion x meets the definition of an instantaneous temporal part (from 7, 8, 9 and 10).
11. Since all diachronic fusions exist (see 6) it follows that every persisting object is composed of these instantaneous objects.
12. So four-dimensionalism is true.

To clarify somewhat, the diachronic fusion x that is an instantaneous object (and, according to Sider, an instantaneous temporal part of persisting y), is a limiting case of a diachronic fusion. Since x is instantaneous, it isn't strictly speaking *diachronic* at all. To render matters a little more intuitive, let us call the limiting case of a diachronic fusion a *synchronic fusion*. Then a synchronic fusion is some instantaneous object that is the fusion of the members of some set S at a time. Sider's argument is essentially that once we admit that composition is unrestricted, we must admit that any arrangement of particulars at some time composes a synchronic fusion. But once we admit that these synchronic fusions exist, then we have admitted that objects are four-dimensional: for persisting objects will be composed of these synchronic fusions.

A similar argument has been proposed by Dowe and Barker.[16] Their argument though, does not rely on a general claim about unrestricted composition. Rather, their argument proceeds only on the assumption that if there exists some persisting object O, then O will exist at multiple regions in space-time. And that seems uncontroversial, whether one is a three-dimensionalist or a four-dimensionalist (or indeed a presentist or eternalist). So, the argument runs, at each of those space-time regions there exists some three-dimensional entity $O_1, O_2, O_3 \ldots O_n$. If unrestricted mereological composition holds, then it follows that we can fuse those entities. Even if composition is restricted, however, Dowe and Barker are confident that we can fuse these *particular* spatio-temporally contiguous entities precisely because they are spatio-temporally contiguous and because (most likely) there exist certain causal relations between each of the entities. Suppose Dowe and Barker are right. Then call the fusion of these entities F. F is a perduring four-dimensional object, for F will have as (temporal) parts, each of the three-dimensional objects $O_1 \ldots O_n$.

What is the relation between persisting object O and perduring F? We have two options. Option one is that O is identical to F, and therefore O is a four-dimensional perduring object. Then four-dimensionalism is vindicated. Option two is that there exists both a perduring object F, and an enduring object O which are distinct. This latter option, however, is problematic on two fronts. First, although it could be argued that O is an everyday object and that all everyday objects endure, whereas F is an odd object which perdures, such a view would still accept that there do exist perduring objects. Not only do there exist perduring objects in other possible worlds, there exist actual perduring objects. Worse still, it seems paradoxical that O and F are distinct, and yet O is identical to F at each time at which they exist. For O is identical to O_1, and is identical to O_2 and to O_3 and so forth. For

O_1, O_2, and O_3 just *are* O at each different space-time region. In which case it seems that O and F are both identical and non-identical.

Couched in the terminology introduced above, Dowe and Barker's worry is that for every persisting object O that exists at t_1, t_2 and t_3, there exists some synchronic fusion of O at t_1, O at t_2, and O at t_3. Call those synchronic fusions S_1, S_2 and S_3. Then we can fuse S_1, S_2 and S_3 to create a diachronic fusion F. Finally we must ask ourselves about the relation between F and O, and that is where the problems discussed above arise.

If these arguments are successful then it is surprising. We might have expected that the issue of when composition occurs, and thus *which* objects exist, is orthogonal to the question of *how* objects persist. So let us proceed to examine these arguments more carefully.

3. ENDURING SIMPLES, INSTANTANEOUS OBJECTS

First, let us grant part I of the argument from vagueness. Grant that composition is unrestricted. Does it follow that Sider, Dowe and Barker are right to conclude that for every persisting object O, there will exist a synchronic fusion of O at t for every time t at which O exists? For Sider, we find this claim in its stronger form in premise (2) as the claim that given unrestricted composition, there will exist the fusion of the members of any arbitrary set S at time t, where x is a fusion of the members of S at t iff every member of S is part of x at t, and each part of x at t overlaps at t some member of S. But does unrestricted composition guarantee any such thing? I do not see that the three-dimensionalist need accept that it does.

Unrestricted mereological composition tells us that given that we have some particulars, we can fuse those particulars. That is, it tells us that given that we have some number of *distinct whole* particulars, we can fuse those particulars. Or perhaps we *can* fuse non-distinct (numerically identical) particulars, but if I fuse some object with itself, I get one and the same object. I don't create some further object that has that object as a part. Nor can we fuse 'non-whole' particulars. Suppose there exists some spatially extended mereological simple *MS*. *MS* exists at locations L_1 and at L_2. Does it follow that there exists some fusion of *MS* at a location, in an analogous way to the proposal that there exists some fusion of the members of set S at a time? Does there exist the fusion of *MS* at L_1, and the fusion of *MS* at L_2—call the former *MS*-at-L_1, and the latter *MS*-at-L_2. And does there then exist a further fusion of *MS*-at-L_1 and *MS*-at-L_2? Clearly not. If there existed a fusion of *MS*-at-L_1 and *MS*-at-L_2, that object would be a spatially extended object with each of *MS*-at-L_1 and *MS*-at-L_2 as proper parts. But by definition *MS* is mereologically simple. So unless we want to say that there exists both *MS* and the fusion of *MS*-at-L_1 and *MS*-at-L_2—call it *MSF*—each of which exist at all of the same space-time points, (surely a very unattractive view) then we must reject the idea that we can fuse *MS* at L_1 or at L_2. We cannot fuse *MS* at L_1 because there simply is no whole object that exists at L_2 that is a candidate for being fused.

Similarly, unrestricted composition does not entail that there exist any *instantaneous* particulars that can be fused to compose an instantaneous object. So does it guarantee that we can fuse the members of *S* at *t*? No. We have no reason to suppose that any members of *S* are instantaneous. Suppose *S* has three members, particulars *P*, P_1 and P_2 which persist through *T*. We cannot assume these particulars perdure (after all, this is supposed to be an argument for four-dimensionalism). We cannot assume that *P* is *itself* the fusion of temporal parts *P-at-t*, *P-at-*t_1 and *P-at-*t_2. But unless each of these distinct temporal parts exist, there is no whole object *P-at-t* that is distinct from *P* at t_1 and *P* at t_2. If *P* endures, then the fusion of *P* with itself just is *P*: an enduring object. So while unrestricted mereological composition tells us that we can fuse the (concrete) members of any arbitrary set, and can thus fuse the members of *S*, it does not tell us that we can fuse the members of *S at t* to form some instantaneous synchronic fusion. We can only draw that conclusion if we already know that the members of *S* are instantaneous or perdure.

The problem with Sider's way of talking (in 2) of the fusion of the members of S *at t* is that (2) is ambiguous between the claim that *x* is a fusion-at-t of the members of *S*, and that *x* is a fusion of the members of *S*-at-*t*. Construed as a fusion-at-t, (2) involves some additional mereological axiom that allows for the fusion-at-a-time of particulars which exist at other times, and it is difficult to see why the three-dimensionalist need accept such an axiom. Construed as a fusion of the members of *S*-at-*t*, (2) looks like a tricky way of smuggling in temporal parts by the back door. After all, what is *S*-at-*t* if not a temporal part?

So a reasonable case can be made by the three-dimensionalist for denying that there exist the various synchronic fusions that overlap persisting objects at times, which Sider, Dowe and Barker appeal to. This is true even if one accepts unrestricted mereological composition. Call a version of three-dimensionalism that denies that such synchronic fusions exist *unitary three-dimensionalism.* On this view persisting objects are unitary because there do not exist any instantaneous objects that overlap those persisting objects at times, there is just a single persisting whole. Unitary three-dimensionalism is the most common version of three-dimensionalism, and it is a version that is consistent with either accepting or rejecting unrestricted composition. While some of the rest of this chapter will broadly outline unitary versions of three-, and indeed, four-dimensionalism, a more detailed consideration of these views must wait until chapter five. For most of this chapter will outline and evaluate views that are not unitary—the non-unitary views. But first I return to consider unitary three-dimensionalism.

Notice that embracing unitary three-dimensionalism involves adopting a particular view about the nature of the fundamental furniture of the world. Now, in what follows I will assume either that objects are composed entirely of simples—that composite objects are decomposable into simples without remainder—or they are composed entirely of atomless gunk, where an individual counts as a piece of atomless gunk iff all of its parts have further proper parts. Call the former view atomism, and the latter view the gunk composition view (GC). For most of this book I will assume that atomism is true. This is because, as I will argue, whether one adopts atomism or GC will be largely irrelevant to one's overall view about composition. That is, for any atomistic view, there is a GC view that is strictly analogous, and if some other set of metaphysical commitments militate in favour of a particular view of atomism, then they will equally militate in favour of an analogous view of GC. So for clarity of exposition I will tend to focus on the atomistic case.

So far I have argued that the Sider, Dowe and Barker arguments do not show that unrestricted composition is incompatible with three-dimensionalism, for they do not show that there exist any instantaneous particulars that can be fused to create synchronic fusions. But this is only so given that the three-dimensionalist embraces a particular view about the nature of simples. For clearly if simples were, for instance, point-sized objects in space and time, then the universalist three-dimensionalist would be in trouble. Consider a bunch of simples, all of which exist at t_1, and call the synchronic fusion of those simples S_1. And suppose there exists another synchronic fusion—S_2—at t_2. If mereological universalism is true, then there exists some object—D_1—that is the fusion of F_1 and F_2. Mereological universalism guarantees that D_1 exists, and D_1 is a perduring object, since it has F_1 and F_2 as maximal temporal parts. But one need not accept a claim as strong as universalism to get the same result here: one need only accept that all composition is mereological. Then in a world of point-sized simples, if there exist *any* persisting objects at all, they will be diachronic fusions. So all persisting objects perdure, there are just fewer of them than there would have been had universalism been true.

What this tells us is that if we want to rule out the existence of synchronic fusions and hence such perduring objects, then given that we are atomists, we must assume that the most basic simples themselves persist. Thus the unitary three-dimensionalist will hold that simples endure.

But suppose that instead of being an atomist, one adopted the gunk composition thesis. What would unitary three-dimensionalism look like then? Well, suppose there is some portion of gunk that exists at, and only at, t_1. And suppose there is some other portion of gunk that exists at, and only at t_2. Then, given universalism, there is some fusion of these two portions of gunk, which has them each as proper parts. And that thing is a perduring object. Why think that there is any portion of gunk that exists at and only at, say, t_1? Why think that there *are* any instantaneous portions of gunk? Well, think of it another way. All persisting objects, whether they endure or perdure, occupy some four-dimensional region of space-time. The point of dispute is just: what is the relationship between each of the three-dimensional 'slices'[17] of that volume—is the relation one of identity, or one of part to whole? Suppose some four-dimensional volume is composed of gunk: that is, the gunk is smeared over a four-dimensional region. Call the persisting object that occupies that region *O*. Then we might be tempted to say that there need not by any instantaneous objects—instantaneous portions of gunk—that are parts of *O*. But notice that in order to say this, we have to deny that every extended sub-region of the gunk is some proper part of *O*. We have to say that while every part of the gunk has some proper part, (otherwise it would not be gunk) we cannot divide the gunk into parts just 'any old how'. Rather, any unitary three-dimensionalist must maintain that gunk is synchronically divisible, but not diachronically divisible. That is to say, at any time, the gunk can be divided into spatial parts each of which have further proper spatial parts: but these spatial parts are not diachronically divisible: they are temporally extended. For clearly, if a four-dimensional volume of gunk was diachronically divisible, then *O* would straightforwardly have temporal parts. The point is that we cannot divide the gunk up into instantaneous portions. Instead, we have to think of the proper parts of the gunk as themselves persisting, and presumably for the three-dimensionalist, we are to think of them as enduring. So we might think of these enduring parts of the gunk as *streams* of gunk. Then all of the proper parts of the gunk are long streams of gunk that are synchronically divisible into further streams of gunk which differ only in their spatial extent.

Notice something here. Except for the fact that these streams are always further divisible into proper parts, they look a lot like the enduring simples we just encountered. We will return to consider the GC thesis shortly. For now, however, it should be fairly clear that a unitary version of three-dimensionalism will look very similar regardless of whether one adopts atomism or GC. For as soon as one rejects the existence of instantaneous objects that coincide with persisting objects at times, one is forced to embrace certain views about the fundamental constituents of the world. One is forced to say that if there are simples, then these simples must endure, and if there is gunk, then this gunk is not diachronically divisible, but rather, persisting objects are composed of enduring gunk streams.

But now, notice something else. Just as unitary three-dimensionalists reject the existence of synchronic fusions that overlap persisting objects at times, so too there is an analogous version of four-dimensionalism that rejects the existence of such objects. *Unitary four-dimensionalism* is the view that although persisting objects are four-dimensional and hence extended in time, they are not the fusion of (maximal) temporal parts.[18] For unitary four-dimensionalists, persisting objects are temporally extended simples: they do not wholly exist whenever they are present, but neither are they present whenever they exist in virtue of having some temporal part present at that time. Unitary four-dimensionalism is the view that I introduced in chapter three according to which four-dimensional objects terdure.

In what follows I will discuss unitary three-dimensionalism in more detail—in particular I will discuss how to make sense of restricted and unrestricted composition in terms of unitary three-dimensionalism. Notice however, that essentially the same issues arise when we consider unitary four-dimensionalism. Exactly how to understand restricted and unrestricted versions of unitary four-dimensionalism is a matter that will be taken up further in chapter five. For now we need only develop a broad characterisation of these four-dimensional views.

Notice that although the unitary four-dimensionalist holds that persisting objects are temporal simples, she will not deny that such objects have spatial parts at times: she denies only that there exist any maximal *temporal* parts. Intuitively, the unrestricted unitary four-dimensionalist thinks that for any arrangement of particulars at any times, there is some persisting object that has those particulars as spatial parts at those times. The unrestricted unitary four-dimensionalist thinks that there exists a plethora of four-dimensional objects each of which have different spatial parts at different times. For instance, on this view there exists at *t* some object that has all of the spatial parts of a trout at *t* and a turkey at *t*, even though that object has no temporal part at *t* that is the fusion of trout and turkey parts at *t*. The restricted unitary four-dimensionalist, on the other hand, does not think that for any arbitrary arrangement of particulars at times, there exists some object that has those particulars as spatial parts at those times. Rather, she thinks that only some particulars at times are the spatial parts at those times, of some persisting four-dimensional object.

This preliminary characterisation leaves much that is unexplained. It does not tell us, for instance, how the particulars in question get to be spatial parts of the four-dimensional objects. For the four-dimensional objects are not fusions of temporal parts which in turn are fusions of those spatial parts at times. Nor is it easy to see how these temporal simples could straightforwardly be fusions of persisting temporally extended simples, for reasons analogous to the ones I discuss in the following section. These are issues that I will return to in chapter five. For now I return to the issue of whether unitary three-dimensionalism is a viable view, and if it isn't, whether this entails that persisting objects are four-dimensional and perdure.

3.1 Fusions-at-a-time and Fusions (at a time)

In the previous section I suggested that three-dimensionalists need not accept the existence of Siderian fusions-at-a-time—instantaneous objects that are the fusions, at a time t, of the concrete members of a set S some or all of whose members may exist at times other than t. Might there, however, be arguments in favour of the existence of these fusions-at-a-time? I think there are such arguments, although I do not think they are definitive and thus do not think there is reason to reject unitary three-dimensionalism outright. I do think though, that there are good reasons to prefer a different version of three-dimensionalism. So what are these arguments?

Let us begin, for illustrative purposes, by conceiving of a world w in which there exist just three enduring simples A, B and C. Suppose that A endures from t_0 to t_5, B endures from t_1 to t_6, and C endures from t_2 to t_7. Suppose further that composition is unrestricted. In w then, there exist four fusions: *AB, AC, CB*, and *ABC*. (Notice that the fusion of A and B exists from t_0 to t_6, and thus there are times when only one part of the fusion exists, namely t_0 at which only A exists, and t_6 at which only B exists.) Suppose these fusions endure. Call such objects *enduring fusions*.

Whereas the three-dimensionalist talks of enduring objects having spatial parts at times, it is natural to talk of these fusions as having parts tenselessly. It is tenselessly true of the fusion *AB*, that B is part of *AB*. But the three-dimensionalist will want to say, for instance, that at t_0, the fusion *AB* has A as a spatial part, and that at t_1 the fusion *AB* has both A and B as spatial parts. So, it might be argued, this is good reason to suppose that we do need the locution of a 'fusion-at-a-time', and thus good reason to suppose that unitary three-dimensionalism is implicitly committed to the existence of such objects. In that case there is really no 'unitary' three-dimensionalism: no view that rejects the existence of instantaneous objects that overlap persisting objects at times.

Not so. We need to be careful to distinguish between a 'fusion-at-a-time'—an instantaneous object that is the synchronic fusion of enduring particulars at-a-time—and a 'fusion, at a time'—a fusion *simpliciter* considered at a time. In the latter case, we are considering a fusion *simpliciter*, (an enduring object) and asking which parts it has at some time, we are not implying that there exists any instantaneous object that is a fusion-at-a-time. Thus the fusion of *ABC* tenselessly has enduring objects *A*, *B* and *C* as parts. However, at t_7 only *C* exists. Thus considered at t_7 the fusion of *ABC* has only one spatial part, *C*.

But wait. It seems that unitary three-dimensionalism is faced with some additional problems. Suppose that in *w* there is some object, call it *O*, that exists between t_2 and t_5. Suppose that *O* has spatial parts *A* and *B* at t_2, *B* and *C* at t_3, *A* and *C* at t_4, and *A*, *B* and *C* at t_5. *O* is a perfectly ordinary persisting object in that it exists at different times and has different spatial parts at those times. Notice though, that *O* is not a fusion of enduring simples. It has as spatial parts *A*, *B* and *C*, but it is not the *fusion* of *A*, *B* and *C*, for the fusion of *A*, *B* and *C* is an object that exists between t_0 and t_7, and which considered at each time between t_2 and t_5, has as spatial parts *A*, *B* and *C* at *each* of those times

Why think that objects such as *O* exist? Well, those who are attracted by the core idea behind mereological universalism—that any way of putting together particulars will create some further particular that has those particulars as parts—should surely find it compelling that *O* exists, since *O* is just an object that is composed, at different times, of different simples. But even those who think that composition only occurs under certain circumstances, should think that objects *like* O_1 exist in the actual world.

In the actual world, objects lose parts. Let us distinguish two different senses in which an object might lose a part. An object *O* might lost a part in virtue of having some part *P* at one time *t*, and not having part *P* at a time *t**, in virtue of *P* ceasing to exist at *t**. Call this the weak sense of losing a part. Then a persisting object *O* loses a part *P* in this weak sense just if (i) *P* is part of *O* at all times at which *P* exists and (ii) there is some time *t* at which *O* exists, and *P* does not. Alternatively, an object might lose a part *P* in virtue of *P* ceasing to be part of *O*, despite the fact that *P* continues to persist. Call this the strong sense of losing a part. Then a persisting object *O* loses a part *P* in a strong sense just if (i) there is some time *t* at which *P* is part of *O*, and (ii) there is some time *t** at which both *P* and *O* exist, and (iii) at *t***P* is not part of *O*. It is certainly the case that actual composite objects lose composite parts in the strong sense—imagine me having my legs chopped off by a combine harvester. It is also plausible to think that if simples persist, then actual objects 'strongly' lose simples across time—it would be fortuitous indeed if at every instant at which composite objects change, they do so in virtue of the cessation of existence (and/or the coming into existence) of one or more of the simples of which they are composed. Plausibly then, many actual composite persisting objects are like *O* in that they strongly lose parts. But any object that strongly loses parts cannot be a fusion of those parts. So we have good reason to suppose that 'non-fusions' like O exist if simples endure.

Then call any composite enduring object that is not a fusion of enduring simples an *enduring non-fusion*. The question is, what is the relation between O and the various fusions that exist in w?[19]

Once again, we might think that here we find good reason to embrace the fusion-at-a-time. We might suppose that at each time at which O exists, it is composed of a distinct fusion-at-a-time: at t_2 it is composed of the fusion-at-a-time of A and B at t_2, and the fusion-at-a-time of B and C at t_3 and so forth at the other times. In order to explain the relationship between O and the enduring simples in w, it might seem that the three-dimensionalist needs to admit that there exist fusions-at-times. If that is so, then the fusion of these synchronic fusions-at-times will be a diachronic fusion. So it looks as though it will turn out that O perdures: it is composed of distinct synchronic fusions at each time at which it exists.

This is premature. We already know that three-dimensionalists are committed to the existence of a relation that holds between materially coincident enduring objects at times: the constitution relation. So when we ask what relation holds between O and the various fusions in w, it is consistent that at each time, a relation of constitution holds. The three-dimensionalist can say that at t_2, O is constituted by enduring fusion AB, and at t_3 is constituted by enduring fusion BC, and at t_4 is constituted by enduring fusion AC and so forth. Hence we might say that O is constituted at times, by different enduring fusions. (Thus one and the same enduring fusion may constitute different enduring non-fusions at different times). Just as we were able to talk of the parts of an enduring fusion at a time without invoking some instantaneous object that exists at and only at that time, so too we can talk of an object being constituted by some enduring fusion at a time, without being committed to the existence of some instantaneous fusion-at-a-time that exists at and only at that time.

3.1.1 The Constitution Relation

This is all well and good, you might think, but there is a problem. In chapter one I defined constitution in a preliminary manner as the symmetrical relation that holds at a time t between any two enduring objects O and O^* just when those objects are materially coincident at t. Thus we get to say that O and enduring fusion AB are related by constitution at t_2, and O and enduring fusion BC are related by constitution at t_3 and so forth. Plausibly, however, what it is for some object O to have a part P, is for P to be fused by O. After all, parthood is defined by the axioms of mereology, which tell us that what it is to be an object with parts is to be a fusion of particulars. Since our enduring object O is by definition not a fusion of enduring simples, this raises the question of how that object has parts at all. An obvious answer would be to say that O has spatial parts A and B at t_2 in virtue of being constituted, at t_2, by the fusion AB which does have A and B as parts. Then enduring non-fusions 'inherit' parts from the fusions that constitute them at times. An enduring non-fusion O has a part P at t just if at t, O is constituted by some fusion F that has P as a part at t.

That we can explain how non-fusions have parts at times by appeal to their being constituted at those times by fusions, is an attractive idea. Given our current definition of constitution in terms of material coincidence, however, as it stands this would clearly be circular. So if we want to appeal to the constitution relation to explain in what sense enduring non-fusions have spatial parts, then we need to define constitution in a way that does not itself appeal to parthood. And this is no easy task. Incidentally, it also means that the constitution relation will not be a symmetric relation. Fusions will be the more 'basic' objects, which constitute at times non-fusions—fusions straightforwardly have parts in virtue of being fusions, while non-fusions have parts in virtue of being constituted by fusions at times. Notice though, that this asymmetry does not in any way track the sorts of asymmetries that three-dimensionalists typically attempt to capture when defining constitution. This asymmetry does not, for instance, track the intuitive asymmetry in the relation between a person and a body, a statue and a lump of clay and so forth. To the extent that we see asymmetries in these kinds of cases, those asymmetries are irrelevant to the asymmetric notion of constitution we are here exploring.

So how are we to define constitution? One thought might be that fusions constitute non-fusions at times just when the fusion and the non-fusion share the same spatial extent, or spatially coincide (where to spatially coincide is not simply to exist within the same spatial boundary, but to exist at all of the same spatial points). Unfortunately this will not do. Physicists tell us that not only is it possible, but in fact it may actually be the case that there are particles that can meet and 'inter-penetrate', that is, that can temporarily exist at all of the same spatial points.[20] Yet at the time at which this occurs, we do not want to say that these particles constitute one another. The relation between such particles is different from the relation that holds between a fusion and the non-fusion it constitutes at a time. In this latter case, had the fusion not existed at the relevant time, then the non-fusion would not have existed at that time either. What it is for a non-fusion to exist at a time, is to be constituted by a fusion at that time. Not so for inter-penetrable particles.

Suppose particle P and P^* are inter-penetrable, and they spatially coincide at time t and location L. Then had P not existed at t and L, P^* would still have existed at that time and location.[21] The existence of P is completely independent from the existence of P^*, it is merely a contingent fact that at some time they spatially coincide.

This suggests we might say that a fusion constitutes a non-fusion at a time if the existence of that fusion logically entails the existence of the non-fusion at that time. Hence:

> Constitution: A fusion F constitutes an enduring non-fusion O at a time t just if (i) F and O exist at t, and (ii) the existence of F entails the existence of O at t and (iii) there is no proper part of F whose existence entails the existence of O at t.[2223]

This definition captures the sense in which the existence of a constituted non-fusion at a time depends on the existence of a fusion at that time. Notice that clause (iii) is required in order to prevent, for instance, fusion AB constituting at t_2 some enduring non-fusion O^* that has spatial part A as an *improper* part at t_2.

This definition does require that one embrace a particular view about necessary connections between objects. It requires that one embrace a sort of modified Humeanism about necessary connections between distinct existences. Unlike straight Humeanism, it allows that such connections exist: they exist between constituting and constituted objects. But on this view such connections can *only* exist between constituting and constituted objects. For we do not want it to be the case that there could exist some necessary connection between a fusion O_1 at t_1 on one side of the world, and a non-fusion O_2 at t_1 on the other side of the world. Then by our definition of constitution, O_1 would constitute O_2 at t_1 even though O_1 and O_2 do not spatially coincide.

We could introduce an extra clause into the definition, to ensure that a fusion F constitutes a non-fusion O at t only if at t, F and O spatially coincide. This rules out O_1 from constituting O_2, but it does so in a merely contingent manner. For we may suppose that O_1 is a fusion of the sort of peculiar 'inter-penetrable' particles we discussed earlier, and non-fusion O_2has as spatial parts at times, the same odd inter-penetrable particles. We can also suppose that at some time—t_2 say—O_1 and O_2 come to spatially coincide. Then even by the amended definition, O_1 would constitute O_2 at t_2. That is not what we want. The reason O_1 does not constitute O_2 at any of the other times at which they exist is not because they fail to spatially coincide at those times, but rather, is due to something about the nature of O_1 and O_2 themselves. Namely, the reason we think that O_1 does not constitute O_2 even when they spatially coincide is because we think they are genuinely distinct and independent. Contra our initial supposition, we do not think that O_1 entails O_2, for we do not think that there exists any necessary connection between O_1 and O_2. If such a connection existed then O_1 would constitute O_2. This tells us is that the genuine cases of constitution that we have been considering exhaust all and only the necessary connections between distinct objects.

Now, I do not contend that this is an entirely satisfactory account of constitution. As just noted, it requires that one adopt a particular view about necessary connections between objects. It also leaves it rather perplexing why the existence of a fusion at some time *sometimes* entails that some enduring no-fusion exists at that time, and sometimes (at least on some views) it does not so entail. That is, we know that a non-fusion exists at a time just in virtue of being constituted by some fusion at that time: but why do fusions sometimes constitute non-fusions at times, and sometimes not? But I think this is the best we can do given the constraints. Moreover, as I will argue, any problems with this account of constitution will equally be problems for analogous four-dimensionalist accounts. So I leave them aside.

So far we have defined the relation that holds between fusions and the enduring non-fusions that they constitute at times. But what of the lump and statue? What are we to say about the relation between them? Lumps and statues gain and lose parts. Plausibly, lumps and statues—and most everyday objects—strongly gain and lose parts. Therefore these objects are *both* non-fusions, and cannot, as traditional three-dimensionalists usually hold, be related by constitution as I have defined it. So much of the three-dimensionalist project appears to be in jeopardy.

But all is not lost. We can say that for any time at which the statue and the lump materially coincide, they do so in virtue of being constituted at that time, by the same fusion. The statue and the lump are not directly related by constitution, rather, they are related indirectly in virtue of both being related to some third object: the fusion. Call the relation that holds between two or more enduring non-fusions at a time the relation of being co-constituted. Then:

> Co-constitution: For any two or more enduring non-fusions O and O^* that exist at a time t, O and O^* are co-constituted at t just if (i) O is constituted by some fusion F at t, and (ii) O^* is constituted by some fusion F^* at t, and (iii) F is identical to F^*.

It is the easy to see how the work that was previously done by constitution is now done by co-constitution. We get to say that the lump and the statue are materially coincident at certain times because they are constituted by the same fusion and thus 'inherit' the parts of the fusion that exist at that time. This too explains why the lump and the statue share the same intrinsic properties at those times—they are constituted by the same fusion—and also why they are distinct—there might be times at which the lump is constituted by a different fusion from the statue, and thus times at which they are not co-constituted. These issues will be discussed further in chapter five, for now we need only grasp the general idea of the co-constitution relation.

So far then, we have developed the view that I call unitary three-dimensionalism. Notice that in some ways this is a rather unusual view. It commits us to the existence of both fusions and non-fusions. Moreover, we might have tended to think of ordinary objects as the paradigm objects, and odd gerrymandered objects as merely fusions of those ordinary objects at times and places. For the unitary three-dimensionalist though, amongst composite objects it is fusions that are ontologically basic: non-fusions exist at times solely in virtue of being related to these fusions by the constitution relation.

And, if simples turn out to be long-lived objects, as well they might, then these fusions too will be long-lived things, which are most of the time massively spatially scattered and not at all object-like. We might expect that there will be various fusions of simples many of which existed at the time of the big bang, most of which were scattered throughout space for millions of years, and which briefly coalesced on earth for a short period before dispersing again into the cosmos. It is when these simples thus coalesce, that we are inclined to say that their fusion constitutes, at a time, some non-fusion. So the picture seems to be one of scattered objects that coalesce briefly into non-fusions at times. Notice that this means that we need a lot of these fusions. The unitary three-dimensionalist might not be a universalist, but regardless of how she restricts composition, we need enough fusions about to constitute the non-fusions whose existence we are committed to. And since normal composite objects lose a lot of parts across time, this suggests they are constituted by a variety of different fusions at different times.

Essentially the same is true if one rejects atomism in favour of GC. Once we have long lived enduring streams of gunk that are only parts of an object for a short period of the time during which they exist, we are left with having to say that composite objects are not fusions of those simples or streams. Just as on the atomistic view, we need to appeal to some relation that holds between ordinary objects, and fusions at a time, so too on the gunk view we need to appeal to a similar relation that holds between objects, and fusions of gunk at a time. If we cannot say that I am the fusion of all of the gunk streams that are ever parts of me, then we must say that at each time at which I exist, I am related in some way to some fusion at a time, of gunk.

Then if I change at every moment I exist, then I am related to a different gunk fusion at every time at which I exist. Then different fusions of gunk streams will constitute me at each time at which I exist.

So consider some person who is composed of gunk. Suppose that at time *t*, the person loses a foot. The foot is a proper part of the person: it is a stream of gunk—it has further streams of gunk as proper parts—each of its toes, for instance. The foot gunk stream ceases to be part of the person when the foot is chopped off, though the gunk stream continues to exist. Indeed, composite objects, especially living ones, are gaining and losing bits of matter all the time. And whenever this is happening, we are gaining new streams of gunk as parts, and losing other streams of gunk as parts. Imagine there is some (spatially small) stream of gunk that two weeks ago was a part of an orange in Columbia, and which prior to that was part of some fertiliser, and prior to that part of a cow, and so forth, and which is now part of me since yesterday I ate the orange. Then today it is true of me that I am composed of streams of gunk which last week existed at distant spatial locations. What makes this the 'same' stream of gunk?. That it is spatio-temporally continuous. But it is not always continuous with me. It used not be part of me, and in the future will not be part of me. Likely this is not the usual view we have in mind when we think of objects being composed of gunk. For on this view, composite objects that lose parts are objects that have gunk streams 'streaming' into and out of them all of the time. So what we know is that these everyday objects are not simply fusions of gunk. The fusion of all the gunk streams that are ever part of me, is not an object that is anything like me, and indeed, nor is the fusion of the streams of gunk that are part of me now. Each of these are massively spatially scattered objects that conceivably exist for some very long period of time, depending on the extent to which streams of gunk are destructible.

So the unitary three-dimensionalist's view about composition and about the nature of simples (or gunk) is different indeed to what we might have thought. Once we adopt certain views about persistence, in particular, about whether there exist instantaneous object that coincide with persisting objects at times, it turns out that we are committed to a number of other metaphysical views: the interrelations between these commitments runs deep. So let us continue in our elucidation of unitary three-dimensionalism.

3.1.2 Restricted and Unrestricted Unitary Three-dimensionalism

I noted earlier that three-dimensionalism is consistent with both restricted and unrestricted composition. So far I have talked generally about unitary three-dimensionalism. At this point it is worth clarifying exactly what an unrestricted and a restricted version of unitary three-dimensionalism amount to. First consider unrestricted unitary three-dimensionalism. We might be tempted to say that this the view according to which (i) for any arbitrary set S of enduring simples, there exists some fusion of the members of that set and (ii) for any arbitrary set S of fusions and any arbitrary times at which those fusions exist, there exists some enduring object that is constituted by those fusions at those times. As I discuss in more detail in the following section, however, most three-dimensionalists probably will not want allow that all simples can be fused. For that would allow that temporally nonoverlapping[24] simples can be fused, which is tantamount to allowing that at least some objects perdure. For suppose that enduring simple O_1 exists from t_1 to t_3, and enduring simple O_2 exists from t_5 to t_8. Then although by definition O_1 and O_2 endure, the fusion of O_1 and O_2 perdures, since O_1 is a (temporal) part of that fusion, as is O_2.

Some three-dimensionalists might be happy to concede that such perduring objects exist. She might say, for instance, that normal persisting objects, an account of whose persistence we are trying to give, are indeed enduring objects. It is only odd gerrymandered objects that perdure. And there is no reason to expect that if such objects exist, that they persist in the same way as ordinary objects. For notice that the sorts of perduring objects we are talking about are precisely the kinds of objects that lack the usual sorts of causal continuities. Such objects do not persist in virtue of causally propagating themselves across time, via some sort of immanent causation: they are merely fusions of 'unremarkably' causally related objects. So the terdurantist might be prepared to say that although such objects exist, persist, and indeed, perdure, there is nothing interesting going on here. Indeed, she could make a stronger case than that. She might claim that although such objects exist, and are indeed temporally extended, we should not rightly say that they persist. For perhaps what it is to *persist*, in some robust sense, is to endure: perduring objects then, do not persist at all.

So let us say that the there-dimensionalist who is prepared to accept the existence of at least *some* perduring objects that are fusions of temporally non-overlapping objects, embraces *weak unitarism.* Then a weak unitarist who in addition embraces unrestricted composition, would hold the following view. Let us call it weak unrestricted unitary three-dimensionalism.

> WUU3D: (i) for any arbitrary set S of enduring simples, there exists some fusion of the members of that set and (ii) for any arbitrary set S of fusions and any arbitrary times at which those fusions exist, there exists some enduring object that is constituted by those fusions at those times.

The weak unrestricted unitary view is pretty straightforward, and for those three-dimensionalists who are prepared to countenance the existence of some perduring objects, it might be an attractive position. There is, of course, a restricted version of weak unitarism,[25] however, it seems unlikely that this would be a popular view. Much of the motivation for restricting composition is to avoid ontological commitment to various gerrymandered objects: objects that do not causally self propagate. So it seems unlikely that one would be prepared to admit the existence of only some of the perduring objects that are fusions of temporally non-overlapping objects, and reject the existence of others. For all of these perduring objects are ones that fail to causally self-propagate—they are punctuate objects. So it seems unlikely that one would both accept weak unitarism, and at the same time restrict composition.

Now, I have said that weak unrestricted unitarism might be a plausible view. In fact, I think that even if a three-dimensionalist was prepared to accept the existence of some perduring objects, there are other reasons to be concerned. While there clearly is a distinction to be drawn between ordinary persisting objects, and odd gerrymandered objects, it is not clear that this distinction is captured by the distinction between enduring objects and perduring ones. Remember, what is supposed to ameliorate the fact that it turns out that some objects perdure, is that those objects are odd gerrymandered ones: ordinary objects nevertheless persist by enduring. But it is surely the case that many odd objects that do not causally self-propagate will count as enduring. While we might think, *prima facie*, that being composed of all and only enduring simples that are temporally no-overlapping entails a failure of causal propagation, it certainly does not follow that being a fusion of simples that temporally overlap entails that there *is* such causal propagation. For many such fusions are very odd objects indeed: objects that are, at many times, massively spatially scattered. So the endure/perdure distinction does not capture the distinction between ordinary objects on the one hand, and gerrymandered ones on the other. So although I think that weak unrestricted unitarism is a coherent view, in what follows I will not, discuss it. It should, however, be clear from the discussion of strong unitary views, how the case for equivalence would be made if this was one's preferred view.

So let us turn to consider strong unitary views. These are views that reject the idea that there exist *any* perduring objects. Thus they are views that deny that there exist any fusions of particulars such that the fusion is a punctuate object: they deny that one can fuse temporally non-overlapping objects. Since strong unitarism is the only version that I consider here, I will henceforth refer to strong unitarism simply as unitarism. Strictly speaking then, all unitary views are restricted. There is, nevertheless, a distinction to be drawn between unitary views that embrace unrestricted composition relative to the set of temporally overlapping objects, and those who do not. The former rule out all and only the fusions of non-overlapping objects, while the latter rule out all these fusions, and some others as well. For simplicity I will label these views 'unrestricted' and 'restricted' respectively, though it should be kept in mind that this is restriction or not, relative to an already existing restriction. Then an unrestricted version of unitary three-dimensionalism is the view that:

> UU3D (i) for any arbitrary set S of temporally overlapping[26] simples, there exists a fusion of the members of S and (ii) for any arbitrary set S of fusions, and any arbitrary times at which those fusions exist, there exists some enduring non-fusion that is constituted by those fusions at those times.

Then restricted unitary three-dimensionalism is the view that there exist at least some fusions of temporally overlapping simples, and that at least some of those fusions constitute enduring objects at times. Notice that this is consistent with its being the case that for every fusion and time at which that fusion exists, there exists some non-fusion that is constituted by that fusion at that time. Given that we have defined constitution as an entailment relation between fusions and non-fusions at times, this is a plausible view and is indeed the view that I usually refer to when I talk of restricted unitary three-dimensionalism. Thus we have the following view:

> RU3D: for some sets $S \ldots . S_n$ of temporally overlapping simples, there exist fusions $F \ldots F_n$ of the members of those sets and (ii) for some fusions $F \ldots F_n$ and times $t \ldots t_n$ at which those fusions exist, there exist enduring non-fusions $O \ldots . O_n$ which are constituted by those fusions at those times.

I contend that unitary three-dimensionalism is defensible, and thus an appeal to fusions-at-a-time is not necessary. Thus Sider, Barker and Dowe's arguments all fail, since they require that we accept that there exist synchronic fusions that overlap persisting objects at times. But those who accept unitary three-dimensionalism need not accept any such thing.

I do think, however, that there are some reasons to find unitary three-dimensionalism perplexing. First, given unitary three-dimensionalism there exist two different sorts of enduring objects. On the one hand we have enduring objects that are fusions of enduring simples. In addition to these fusions there exist enduring objects that are constituted by fusions at times. Moreover, as I noted earlier, given unitary three-dimensionalism we have a picture of the world according to which ontologically basic spatially scattered fusions 'coalesce' briefly and constitute less basic enduring non-fusions. But none of this suggests that unitary three-dimensionalism is incoherent.

There is, however, also another reason we might be tempted to acknowledge that there exist fusions-at-times. I argued previously that only whole objects can be fused, and thus that we can only fuse the members of *S* at *t*, if the members of *S* are themselves instantaneous objects. Thus we can fuse objects to create instantaneous fusions, only so long as there already exist instantaneous objects to fuse. But unrestricted composition does not guarantee that there exist any such instantaneous objects, and it certainly does not guarantee that for every persisting object, there exist instantaneous objects that compose that object at a time.

It might be objected, however, that since three-dimensionalism holds that enduring objects are wholly present whenever they exist, then there is every reason to suppose that we can fuse enduring particulars at times. For contrary to my analogy with the spatially extended mereological simple *MS* which is, presumably, not 'wholly present' at any of the spatial locations at which it exists, an enduring object *is* wholly present at each temporal location at which it exists. So even if some particular *P* that is a member of set *S* endures, and thus exists at t_1, t_2 and t_3, by definition *P* is supposed to wholly exist at each of these times. Yet if all of *P* exists at t_1, (and at t_2 and t_3) then if we can fuse wholes, we ought to be able to fuse *P* at t_1, with some other enduring particular at t_1. So we ought to be able to fuse the members of *S* at *t* regardless of whether those members endure. Hence the three-dimensionalist ought to accept the existence of fusions-at-times, and, if Sider is right, ought then to conclude that persisting objects are four-dimensional.

While this argument does not decisively show that the three-dimensionalist ought to accept the existence of fusions-at-times, it does have intuitive appeal. The question then becomes whether accepting the existence of such synchronic fusions-at-times is accepting four-dimensionalism. In the following section I argue that it is not. In doing so, I develop a non-mereological account of composition. I then go on, in section three, to develop a novel version of three-dimensionalism that embraces the existence of these 'fusions-at-times', a version of three-dimensionalism that I then consider in detail in chapter six.

3.2 Fusions-at-times and Temporal Parts

Let us grant not only that composition is unrestricted, but further, that the three-dimensionalist ought to accept the existence of synchronic fusions-at-times that overlap persisting objects at times. Now let us consider whether Sider's argument establishes that such objects count as temporal parts. The crucial step in the argument is premise 8, which relies on the mereological principle that licenses the move from 'every part of *x* at *t* overlaps *y* at *t*' to '*x* is part of *y* at *t*.'

To clarify this step we need to be a little clearer about the move from premise 7 to 8. Sider's idea is that we begin with some persisting object *y*. At some arbitrary time *t*, *y* is composed of some things. Consider the set *y* whose members are all of those things that compose *y* at *t*. The argument is a little confusing because Sider uses '*y*' to refer both to the persisting object, and to the set whose members compose *y* at *t*. I retain this terminology because ultimately it allows us to see how the argument goes wrong. Given that composition is unrestricted, (and we concede there exist fusions-at-times) we can fuse all of the members of *y* at *t*, and call this fusion *x*. Then we conclude that the instantaneous object *x*, is part of the persisting object *y*. We derive premise 7 from the definition of fusion, combined with the fact that *x* is the fusion of the members of *y* at *t*. If x is the fusion of the members of *y* at *t*, then we know that every member of *y* at *t* is part of *x* at *t*, and *x* at *t* overlaps at *t* some member of *y*. So 7 should more properly read:

7. Every member of *y* is part of *x* at *t* and every part of *x* at *t* overlaps some member of *y* at *t*.

Then in premise 8 we move from 'every part of *x* at *t* overlaps *y* at *t*' to '*x* is part of *y* at *t*. This should instead read: 'every part of *x* at *t* overlaps some *member* of *y* at t.' And it is not obvious that '*x* is part of *y* at *t*' follows from this claim, even if we accept the relevant mereological principle. For presumably the '*y*' in '*x* is part of *y* at *t*', refers to the persisting object *y* at *t*: for only if *x* was part of this object, could it be said that we have shown four-dimensionalism to be true. Moreover, since no fusion is ever part of a *set*, it could not be that *x* is part of *y* at *t*, where '*y* at *t*' is a set at a time. In order to show that *x* is part of *y* at *t*, we need to note that every part of *y* at *t* overlaps some member of the set *y* at *t*, and every member of set *y* at *t* is part of *y* at *t*. And that seems plausible. It *then* follows that *x* is part of *y* at *t*.

Is conceding that *x* is part of *y* at *t*, conceding that *x* is a temporal part of *y*? I say not. We will return to consider this issue in section 4. For now let us put aside the issue of the relation between *x* and *y*, and consider Sider's argument more generally. Does Sider's argument show that given unrestricted composition, it follows that four-dimensionalism is true? If we grant (2), then the argument shows that there exist a plethora of instantaneous synchronic fusions that overlap persisting objects at times. Given unrestricted mereological composition, (or even the weaker principle advocated by Dowe and Barker) it follows that we can fuse each of those temporally contiguous synchronic fusions into a diachronic fusion. Diachronic fusions are four-dimensional objects—for at every time at which they exist, they do so in virtue of some part (a synchronic fusion) existing at that time: hence they persist by perduring.

That conclusion alone does not entail that *y* persists by perduring. It is still open to someone to hold that *y* endures, but that there exists some *additional* four-dimensional object that overlaps *y* at times. It hardly seems likely though, that many three-dimensionalists would embrace such a move. For that would involve accepting that a good many objects do perdure, and that these perduring objects overlap enduring objects in, as Dowe and Barker point out, a perplexing, possibly paradoxical, manner.[27]

At this point we need to return to Part II of Sider's argument. Recall that we assumed in that argument that composition was unrestricted. In fact though, we did more than that. We assumed that the thesis of unrestricted composition should be understood in mereological terms,[28] as the claim that for every set *S* whose members are concrete particulars, there is a mereological fusion of the members of *S*: we assumed that for any concrete particulars, there is a fusion of those particulars. We assumed the view often known as mereological universalism.

But notice something. So long as we think that our world is one in which there exist multiple particulars, at least some of which do not temporally overlap, then mereological universalism *guarantees* that at least some objects perdure. As I noted earlier, it guarantees that if there is some object O that exists from t_1—t_5, and some object O^* that exists from t_7—t_{10}, then there is some third object F that is the fusion of O and O^*. F is a four-dimensional object in that it is only partly present at each time at which it exists, in virtue of one of its parts (O and then O^*), being present at each of those times. Although mereological universalism does not rule out that O and O^* are *themselves* enduring objects and thus wholly present whenever they exist, it turns out that there is some further object F that has these objects as temporal parts. So as long as we think that there exist some temporally non-overlapping objects, as surely all three-dimensionalists do, we do not need a complicated Siderian argument to show that *some* objects perdure:[29] for that is just entailed by mereological universalism. That is why when I defined both restricted and unrestricted versions of unitary three-dimensionalism, I allowed only that temporally overlapping simples could be fused.

Thus even three-dimensionalists who hold only the weak thesis that no *actual* objects perdure, will want to reject the thesis of mereological universalism. Specifically, the three-dimensionalist will want to reject the idea that it is possible to fuse distinct objects that exist at, and *only at* different times. For she wants to deny that there exist any diachronic fusions. And this is hardly surprising. The distinction between fusing objects at a time—synchronic fusions—and fusing objects at different times—diachronic fusions, is not only a principled distinction, but is an obvious one for the three-dimensionalist to make. After all, three-dimensionalists are committed to the claim that there is something substantially different about the way an object persists through time, and the way it extends through space: it is precisely the fact that perdurantists view persistence through time as analogous to extension through space that three-dimensionalists find repugnant. So while the three-dimensionalist *can* accept that there exist various synchronic fusions, she cannot accept that there exist any diachronic fusions. Let us call versions of three-dimensionalism that accept that there exist synchronic fusions-at-times, that is, instantaneous objects that overlap or materially coincide with persisting objects at times, *non-unitary three dimensionalisms*.

Notice then, that the non-unitary three-dimensionalist can embrace any view about the nature of simples: she can hold that they are instantaneous point-sized objects, or that they are spatially and/or temporally extended objects (Though if she holds this latter, she will need to hold that there can exist fusions-at-times of objects that exist at other times). But the non-unitary three-dimensionalist will need to modify her account of composition. If there exist instantaneous objects that coincide with persisting objects at times, then these persisting objects cannot be fusions of those instantaneous objects. Does this mean that the non-unitary three-dimensionalist must reject the thesis of unrestricted composition? No. The general claim that any arbitrary arrangement of particulars at or across time composes some object, need not be understood as the thesis that mereological universalism is true. Rather, the non-unitary three-dimensionalist needs a non-mereological way of cashing out the idea of unrestricted composition. That is, she needs a non-mereological account of across-time composition.

Since it is unrestricted composition that is held to provide the most problems for the three-dimensionalist, or in particular, the non-unitary three-dimensionalist, I begin by developing a universalist version of non-unitary three-dimensionalism. I turn to non-universalist (or restricted compositionalists) versions later. Then let us call a version of unrestricted composition that appeals to a non-mereological relation, *non-mereological universalism.*

Fortunately, the three-dimensionalist already has at her disposal the resources for developing a 'non-mereological universalism'. As we have already noted, the non-unitary three-dimensionalist will distinguish between fusing at a time, and fusing across time. Thus the non-unitary three-dimensionalist who accepts unrestricted composition will follow Sider, Dowe and Barker, in maintaining that for or any arbitrary set S of concrete particulars all of which exist at instant t, there exists some synchronic fusion of the members of S at t. Then what of the relationship between synchronic fusions and persisting objects if persisting objects are not fusions of synchronic fusions: if persisting objects are not just diachronic fusions?

3.2.1 Constitution Revisited

The answer is that the relation must be non-mereological: the relation between synchronic fusions and the persisting objects they materially coincide with at times cannot be the part/whole relation. This brings us neatly back to the constitution relation. For surely the non-unitary three-dimensionalist should say that enduring objects are *constituted* by synchronic fusions at times. Indeed, the non-unitary theorist can avail herself of the definition of constitution we discussed previously:

> Constitution: A fusion F constitutes an enduring non-fusion O at a time t just if (i) F and O exist at t and, (ii) the existence of F entails the existence of O at t and (iii) there is no proper part of F whose existence entails the existence of O at t.[30]

So enduring objects are constituted at different times, by different fusions. So an enduring object is not a fusion of these different fusions: instead, it is non-mereologically related to them. The enduring object changes over time because at different times, it is constituted by different fusions.

This brings us to consideration of the relation that holds between enduring objects that coincide at a time, such as in the case of the statue and the lump. Here too, the non-unitary three-dimensionalist can appeal to the relation of co-constitution in the same way as the unitary theorist. Thus:

> Co-constitution: For any two or more enduring non-fusions O and O^* that exist at a time t, O and O^* are co-constituted at t just if at (i) O is constituted by some fusion F at t, and (ii) O^* is constituted by some fusion F^* at t, and (iii) F is identical to F^*.[31]

Thus any two or more enduring objects, such as the statue and the lump, will be co-constituted at t just if at t, some synchronic fusion F constitutes the statue, and the same synchronic fusion constitutes the lump. This account will be discussed more fully in chapter six. For now we need only show that non-unitary three-dimensionalism is a viable account that is consistent with the view that composition is unrestricted. If we can show this, then we have shown that the existence of synchronic fusions that materially coincide with persisting objects at times, does not entail that those objects perdure.

3.2.2 Restricted and Unrestricted Non-unitary Three-dimensionalism

At this point we need to take care to distinguish two different views that the non-unitary three-dimensionalist might embrace. The first of these is the view that naturally falls out of a consideration of the Dowe and Barker argument. It is the view that whenever we have some persisting object O that exists at times $t_1, t_2 \ldots t_n$, then we have some synchronic fusion-at-a-time of the particulars that compose O at each of those times, that is, we have a fusion of the spatial parts of O at t_1, of O at t_2 and so forth. This is the view that persisting objects are, at every time at which they exist, constituted by some synchronic fusion at each of those times. This is consistent with both the view that composition is restricted, and the view that composition is unrestricted. All it tells us is that *whichever* persisting objects exist, those objects are constituted by synchronic fusions at those times: it does not tell us anything about which objects exist. It tells us only that persisting objects are non-unitary: whenever they exist they are constituted at each of those times, by synchronic fusions. The thesis that persisting objects are non-unitary (the non-unitary thesis) can be expressed as follows:

> NUT(3D): for every persisting object O and time t at which O exists, there exists some synchronic fusion S such that O is constituted by S at t.

There is also a four-dimensionalist version of the non-unitary thesis. The non-unitary four-dimensionalist will agree that for every persisting object O and time t at which O exists, there is some synchronic fusion of the parts of O at t. For the four-dimensionalist though, these synchronic fusions do not constitute O at those times, but rather, are instantaneous temporal parts of O. Non-unitary four-dimensionalism, is, of course, perdurantism by another name.

> NUT(4D): for every persisting object O and time t at which O exists, there exists some synchronic fusion S, such that S is an instantaneous temporal part of O at t.

Just as unitary three and four-dimensionalism come in both restricted and unrestricted versions, the same is true for non-unitary versions of each. As we noted previously, most often perdurantists hold that composition is unrestricted, that is, they are mereological universalists. But not in all cases. There are non-universalist perdurantists such as Storrs McCall, who reject the idea that gerrymandered objects like trout-turkeys exist.[32] On

McCall's view only the 'everyday' objects of our ontology exist, and these objects persist by perduring. This view is a restricted non-unitary four-dimensionalism. Call this view, *restricted perdurantism.* Then the restricted perdurantist agrees with the 'traditional' perdurantist that all persisting objects are mereological fusions of temporal parts, he merely disagrees about which persisting objects exist. Similarly, non-unitary three-dimensionalists all agree that persisting objects are constituted by synchronic fusions at every time at which they exist, they just disagree about which persisting objects exist. Thus a non-unitary three-dimensionalist who rejects unrestricted composition is one who embraces non-unitary restricted three-dimensionalism, or, by parity with the four-dimensionalist case, what we will call *restricted endurantism.*

Finally, there are versions of non-unitary three- and four-dimensionalism that embrace unrestricted composition. This combination of views is probably the most common four-dimensionalist position. It is the view that combines perdurantism with mereological universalism. Call this view *unrestricted perdurantism.* Like restricted perdurantists, the unrestricted perdurantist accepts NUT(4D). But the unrestricted perdurantist thinks that *every* combination of particulars at and across time composes some persisting object. So the unrestricted perdurantist, in addition to accepting NUT(4D), also embraces the thesis of mereological universalism:

> MU: for any arbitrary time t and set S of concrete particulars, there exists a fusion of the members of S at t (a synchronic fusion-at-a-time) and for any arbitrary set S^* of synchronic fusions that exist at distinct times $t_1 \ldots t_n$, there exists a diachronic fusion of the members of S^*.

In essence, unrestricted perdurantism is the view that one can fuse any particulars at a time to create a synchronic fusion, and one can fuse any synchronic fusions to create diachronic fusions. But how are we to understand unrestricted endurantism, the analogous view that combines non-unitary three-dimensionalism with unrestricted composition? To develop that view, we need to develop an account of non-mereological universalism. It is to this that I now turn.

4. UNRESTRICTED ENDURANTISM

Unrestricted endurantism is, intuitively, the view that for any arbitrary temporal interval and arbitrary particulars that exist within that interval, there exists some enduring object O whose temporal extent falls exactly during that interval, and which at each time at which it exists, is composed of those particulars. Moreover, the unrestricted endurantist will be committed to the non-unitary thesis in its three-dimensionalist form. But just as the unrestricted perdurantist has commitments over and above being committed to the non-unitary thesis in its four-dimensionalist guise, so too does the unrestricted endurantist. Unlike the unrestricted perdurantist, however, who understands unrestricted composition in terms of mereological universalism (MU), the unrestricted endurantist must embrace a non-mereological universalism. For the unrestricted endurantist does not hold that persisting objects are fusions of synchronic fusions. Thus we must begin by defining non-mereological universalism. Hence we have the following:

> NMU: for any arbitrary time t and set S of concrete particulars, there exists a fusion of the members of S at t (a synchronic fusion-at-a-time) and for any arbitrary set S^* of synchronic fusions that exist at distinct times $t_1 \ldots t_n$, there exists an enduring object O that is at each of those times, constituted by one of those fusions.

Thus the unrestricted endurantist is one who embraces NMU, essentially the view that for any two or more arbitrary synchronic fusions $F_1, F_2, \ldots F_n$ that exist at times $t_1, t_2 \ldots t_n$ there exists some enduring object O that is constituted by F_1 at t_1, F_2 at t_2, ...F_n at t_n.

To illustrate, let us consider a world w^* in which at t_1 there exist particulars X, Y, and Z, and at t_2 there exist particulars P and Q. Given the first clause of NMU, (and of MU) it follows that there exists at t_1 some synchronic fusion of X, Y and Z at t_1, and that at t_2 there exists some synchronic fusion of P and Q at t_2. Call the former fusion O_1, and the latter O_2. Then the perdurantist, who understands unrestricted composition in terms of mereological universalism, will hold that there exists some diachronic fusion of O_1 and O_2—call it O^*—which is a four-dimensional object that has O_1 and O_2 as (instantaneous) temporal parts. The unrestricted endurantist will deny this claim. Instead, she will hold that there exists some persisting object—call it O—that is constituted by O_1 at t_1, and by O_2 at t_2, and that claim is perfectly consistent with O being a three-dimensional object. For O is wholly present at t_1 in virtue of being constituted at t_1, by some wholly present object, namely O_1, and *mutatis mutandis* for t_2.

Notice that the ontologies of unrestricted endurantism and unrestricted perdurantism are very similar. Both are committed to the existence of the same number of objects, existing for the same duration of time. Both agree that not only are there persisting objects that we call tenors and turnips, but that there also exists a temporally scattered persisting object that wholly overlaps or coincides with, a tenor at t_1 and a turnip at t_7. Call this object a tenor-turnip. For the perdurantist, a tenor-turnip is a four-dimensional object that is the fusion of two instantaneous objects one of which is a temporal part of a tenor, and the other of which is a temporal part of a turnip. For the unrestricted endurantist, a tenor-turnip is an enduring object, which, at t_1 is related by constitution to a synchronic fusion that is also related by constitution at that time to a tenor, and which at t_7 is related by constitution to a synchronic fusion that at t_7 is also related by constitution to a turnip. So on this view, the enduring objects are not all and only the ordinary objects: enduring can be spatially and temporally scattered, they can be punctuate, and they can fail to be self propagating.

The difference between unrestricted endurantism and unrestricted perdurantism is that in the case of persisting objects O and O^*, O_1 and O_2 are proper parts (proper temporal parts) of O^*, but are not proper parts of O. Moreover, given the way three- and four-dimensionalists usually construe the notion of '*simpliciter*', the following will be true: since a four-dimensional object that is the mereological fusion of instantaneous objects is an object that has each of these instantaneous objects as parts *simpliciter*, it is therefore tenselessly true that O_1 is a part of O^*. The three-dimensionalist, however, rejects the idea that persisting objects have parts *simpliciter* in this sense (the sense that does not distinguish between having parts or properties M- or S-*simpliciter*). For the three-dimensionalist, the instantaneous objects that constitute enduring objects at times are not parts *simpliciter* of those objects. Rather, the synchronic fusion O_1 that constitutes enduring object O at t_1, is *at t* an improper part of O, but is not part of O *simpliciter*.

We can further clarify this distinction if we consider how Sider's argument would proceed with respect to extended temporal parts. Consider Daisy*. Daisy* is a diachronic fusion of the fusions of the members of sets C at t, C_1 at t_1, C_2 at t_2 and C_3 at t_3. Let us say that at t, every member of C is part of some other object at t—Daisy—and every part of Daisy at t overlaps some member of C at t. So too with C_1 at t_1 and so forth. Daisy is a cat, and therefore exists at many times at which Daisy* does not. Is Daisy* an extended temporal part of Daisy? To answer this question, consider the following definition, due to Sider, of an extended (maximal) temporal part.

> ETP: x is an extended temporal part of y during T iff (1) x exists at, but only at, times in T (2) x is part of y at every time during T, and (3) at every moment in Tx overlaps everything that is part of y at that moment.[33]

Daisy* exists through times t to t_3. Call this duration T. So Daisy* exists at and only at times in T. Daisy* is part of Daisy at every time during T, and at every moment in T, Daisy* overlaps everything that is part of Daisy at that moment. So we can conclude that Daisy* is a temporal part of Daisy.

All well and good. The three-dimensionalist need not be worried here, since she denies that the diachronic fusion Daisy* exists.

But now let us consider the enduring object Snowy*. Snowy* is an object that is constituted by the fusions of the members of sets D at t, D_1 at t_1, D_2 at t_2 and D_3 at t_3—that is, Snowy* is constituted by the synchronic fusion of the members of D at t, the synchronic fusion of the members of D_1 at t_1 and so forth. At each of those times at which those synchronic fusions constitute Snowy*, they also constitute Snowy, who is a dog (a dog whose temporal extent is considerably greater than a mere four temporal instants). By definition, Snowy* and Snowy are enduring objects that are wholly present whenever they exist: Snowy* is not a temporal part of Snowy. When we look to Sider's definition of an extended temporal part, however, we run into difficulties. For Snowy* exists only during T. Snowy* overlaps at every moment in T, everything that is part of Snowy at that moment, and Snowy* is part of Snowy at every time in T. So by Sider's definition of an extended temporal part, Snowy* is a temporal part of Snowy.

What is Snowy*? Snowy* is simply a persisting object that exists between and only between certain times, and which happens to overlap another object, Snowy, at the times at which it exists. None of this precludes Snowy* (and Snowy) from being wholly present at every time at which each exists. The problem lies in Sider's definition of an extended temporal part. For on this definition, something is an extended temporal part of y if it completely overlaps y for some period of time, and exists only during that period of time. Once we see, however, that the three-dimensionalist can embrace synchronic fusions-at-times in the form of a non-unitary version of three-dimensionalism, then this definition of an extended temporal part is seen to be lacking. For given non-unitary three-dimensionalism, an enduring object can overlap another such object during and only during some period of time T. But nothing about this suggests that one of those objects is a part *simpliciter* of the other. Specifically, nothing about Snowy* shows that it is part of Snowy *simpliciter*.

The problem with Sider's ETP definition of an extended temporal part is that clause (2) is couched in terms of parthood at times, rather than atemporal parthood. Clause (2) does not distinguish between a case where we have an object such as Snowy*, where Snowy* is an improper part of Snowy *at each time t* in T, and the case where we have a diachronic fusion Daisy* that is part of Daisy *simpliciter*. For if Daisy* is part of Daisy *simpliciter* then it is also true that Daisy* is part of Daisy at each time in T, though the reverse is not the case. But only if Daisy* is part of Daisy s*impliciter*, does it follow that Daisy is a temporally extended object, of which Daisy* is an extended temporal part. Sider's atemporal version of ETP achieves just this:

> AETP: x is an extended temporal part of y during T iff (1) x exists at, but only at, times in T (2) x is part of y and (3) at every moment in T, x overlaps every part of y that exists at that moment.

Considering again clause (2), we see that Daisy* is part of Daisy, and thus that Daisy* is an extended temporal part of Daisy. Since both Snowy* and Snowy are wholly present at each time at which they exist, there is no atemporal sense in which Snowy* is part of Snowy. While there are some times at which Snowy* is part of Snowy, there are other times at which Snowy* is not part of Snowy. So (2) is not true of Snowy* and Snowy, and thus Snowy* is not an extended temporal part of Snowy.

This brings us back nicely to the question of why the second part of the argument from vagueness does not show that the fusion of the members of y at t, namely x, is part of the persisting object y. Recall that Sider employed a mereological principle according to which we can move from the claim that every part of x at t overlaps y at t, to the claim that x is part of y at t. I earlier conceded that Sider had shown that x at is part of y at t, where y is a persisting object and x a synchronic fusion. We can now see why conceding this was not conceding that x is a *temporal* part of y. Consider again Snowy* and Snowy. Snowy* at t overlaps Snowy at t. By the mereological principle to which Sider avails himself, we can conclude that Snowy* at t is part of Snowy at t. Even having shown this, however, we have not shown that Snowy* is a temporal part of Snowy. To show that, it needs to be the case that we can move from the claim that Snowy* at t overlaps Snowy at t to the claim that Snowy* is part of Snowy *simpliciter*.

For the non-unitary three-dimensionalist does not deny that at t, Snowy* is part of Snowy: Snowy* is an improper part of Snowy at t. Rather, she denies that Snowy* is part of Snowy *simpliciter*. So while the mereological principle is sound, applying this principle to two materially coincident persisting objects that exist for different durations, tells us only that at each time at which both exist, each is an improper part of the other—that is, they are related by constitution at that time. There is a further question as to whether one is a part of the other *simpliciter*. Only if the answer to this question is 'yes', can we conclude that one is a temporal part of the other. So all Sider's earlier argument shows is that x at t is part of y at t. It does not show that x is part of y *simpliciter*, and thus does not rule out the possibility that y is an enduring object, that is constituted by x at t, rather than a diachronic fusion that has x as a temporal part.

5. WHERE TO NOW?

So far then, we have found that there is a two-way distinction that cuts across the issue of the manner in which objects persist. There is the distinction between a theory that is unitary versus non-unitary, and between a theory that is restricted versus unrestricted. This means that for each theory of persistence there are four possible versions of that theory that differently combine the unitary versus non-unitary, and restricted versus unrestricted distinctions.

The interest in these theories is twofold. First, consideration of these theories tells us something interesting about the relation between composition and persistence. It tells us what sorts of metaphysical packages are coherent, and what sorts of ancillary metaphysical commitments one must embrace if one is to be committed to certain other views about persistence. For instance, it tells us that if you are a three-dimensionalist and you want to deny that there exist a plethora of instantaneous objects that coincide with persisting objects at times, then there are various others views that you have to adopt: you have to adopt a particular view about the nature of simples; you have to adopt a certain view about restrictions on composition; you have to embrace the existence of a certain relation that holds between objects at times; you have to concede that there exist multiple kinds of persisting object, and so forth. It tells us that if you are a three-dimensionalist and you want to accept that there exist these instantaneous objects, then you must reject a mereological conception of composition across time in favour of a non-mereological account, and you must embrace some relation that holds between coinciding objects at times.

Second, a number of these theories are novel ones. This offers the possibility of fashioning a new account of persistence that may be at least as, or perhaps more, plausible than the traditional accounts. Thus these different accounts of persistence may preserve rather different intuitions about objects and their persistence, and thus provide different explanatory apparatus for dealing with the various puzzles we have encountered. They may therefore provide plausible alternatives to either of the current theories. Indeed, in chapter six I consider in more detail the view I have called non-unitary three-dimensionalism. And I argue that this view has much to recommend it over traditional three-dimensionalism.

But suppose that one does not find these new accounts of persistence plausible. Then what is interesting about these theories is what they tell us about why traditional versions of three- and/or four-dimensionalism are successful. For consideration of unsuccessful versions of, say three-dimensionalism, reveal something about which features of traditional three-dimensionalism are essential to that theory: that is, which features you simply cannot change and still have a successful theory.

Finally, sorting out these different versions of three- and four-dimensionalism is essential to the project of determining whether three- and four-dimensionalism are metaphysically equivalent. For, I argue, we must consider analogous versions of each: we must consider versions of each that accept the same ancillary metaphysical commitments. So, for instance, we should compare non-unitary versions of each theory, rather than comparing a non-unitary version of three-dimensionalism, with a unitary version of four-dimensionalism. Then if we can show that for every version of three-dimensionalism there is an analogous version of four-dimensionalism that is metaphysically equivalent and *vice versa*, we have shown that three- and four-dimensionalism are metaphysically equivalent *simpliciter*. In fact although I do consider a number of analogous versions of three- and four-dimensionalism in the following two chapters, I do not consider every version: space and time constraints simply do not allow for that. I do think, however, that all of the most plausible and important versions of each theory are considered, and that if we can show that analogous versions of each of these theories are equivalent, then this is good reason to suppose that three- and four-dimensionalism more broadly construed are equivalent.

NOTES

[1] For instance Van Inwagen (1990) and Merricks (2000).
[2] Sider (2001).
[3] Lewis (1986) pp 212–213. and Lewis 1991. pp 80–81.
[4] Lewis (1986) pp 212–213.
[5] For discussion of these issues see Evans (1978); Hyde (1998); Lewis (1988); Sainsbury (1989) and Tye (1990).
[6] Sider (2001) pp 120–140 and Sider (2003).
[7] See for instance Koslicki (2003). Other advocates of the view that composition is not unrestricted, and vagueness is ontological include Van Inwagen (1990).
[8] Sider (2003).
[9] Sider (2001) pp 134–149 and Sider (2003). I use the terminology from the latter, which differ slightly from that used in the former.
[10] Sider (2003) footnote 2.
[11] Sider (2003) p 135.
[12] Sider (2003) p 135.
[13] Sider (2003) p 136 (x and y are transposed in this chapter: Sider uses 'x' to refer to the set, and 'y' to refer to the fusion of the members of x).
[14] Sider (2003). See footnote 4. p 136.
[15] Sider (2001) p 60.
[16] Dowe and Barker (2003).

[17] Where a slice is not (necessarily) a part of the volume, but just refers to the volume considered at a time.
[18] Where a maximal temporal part is a part that overlaps all of the spatial parts at a time, of the four-dimensional object of which it is a part.
[19] Notice that this question would arise even if we rejected mereological universalism in favour of restricted mereological composition so long as there are some fusions of enduring simples, and some enduring objects that are not fusions but which overlap fusions at times.
[20] Potemkin (2004).
[21] Actually this is not quite right. Even if we set aside worries about deviant causal chains, if it turns out that P and P^* have some sort of charge, then it might be the case that had P not existed at the relevant time and location, then although P^* would still have existed, it would have existed at a slightly different location. In fact, this might even be nomologically necessary. Then there is no robust sense in which had P not existed at L at t, then P^* would still have existed at L at t. Then to capture the sense in which P and P^* are independent, we could say something like the following: P and P^* are independent and distinct just if there is some world that is identical to our world up to time t, such that when at t, God makes it the case that P does not exist, it is still the case that P^* exists at t at L. The idea here is just that at t, God could make it the case that P does not exist, yet P^* would still exist at the same location. I thank David Braddon-Mitchell for this suggestion..
[22] Henceforth I will sometimes talk of F entailing the existence of O, or F entailing O, by which I just mean that the existence of F entails the existence of O.
[23] There is another potential worry here. As it stands it seems that objects and their unit sets might count as standing in a constitution relation. Since plausibly, the existence of some object, O, entails the existence of the set $\{O\}$, then we should say that O constitutes $\{O\}$. Now, this might not make everyone uncomfortable; perhaps this is precisely what we should say about sets. But plausibly, many of us think that $\{O\}$ is an abstract object, and is not the right sort of thing to be constituted by O. Then we can simply amend the definition to include a fourth clause (iv) for any two objects F and O, F constitutes O only if, if F is a concrete particular, then so is O.
[24] Where particulars P and P^* are temporally nonoverlapping just in case there is no time t at which both P and P^* exist.
[25] SRU3D: for some sets $S....S_n$ of simples, there exist fusions $F...F_n$ of the members of those sets and (ii) for some fusions $F...F_n$ and times $t...t_n$ at which those fusions exist, there exist enduring non-fusions $O....O_n$ which are constituted by those fusions at those times.
[26] Where any two particulars $P$and P^* are temporally overlapping just if there is some time t at which both P and P^* exist.
[27] Dowe and Barker (2003).
[28] Lewis (1991); Sider (2003).
[29] Depending on exactly what we mean by 'perdure'. We could decide that something perdures only if it at every time at which it exists, it has some instantaneous temporal part at that time. Then these objects do not perdure. On the other hand, they don't really endure either. I take it that an object perdures just if it has some maximal temporal parts. Then these objects perdure.
[30] This definition refers to enduring non-fusions. In the context of unitary three-dimensionalism, enduring objects can either be fusions or non-fusions and it is necessary to distinguish between the two. In the context of non-unitary three-dimensionalism all fusions

are synchronic, and all enduring objects are non-fusions, so talk of 'enduring non-fusions' is tautologous.

[31] Once again, where for the unitary theorist fusions *F* and F* may be enduring fusions, for the non-unitary theorist *F* and *F** will be synchronic fusions.

[32] McCall (1994) pp. 211–214.

[33] Sider (2001). This definition is also accepted by Markosian (1994).

Chapter 5

THE METAPHYSICAL EQUIVALENCE OF UNITARY THREE- AND FOUR-DIMENSIONALISM

In this chapter I begin the work of showing that three- and four-dimensionalism are metaphysically equivalent. This does not mean that any combination of views that embraces three-dimensionalism, is equivalent to any combination of views that embraces four-dimensionalism. A three-dimensionalist presentist who believes in restricted composition has a view substantially different to a four-dimensionalist eternalist who believes in unrestricted composition (unless it turns out that presentism and eternalism are equivalent as are restricted and unrestricted composition). I will argue that for any *bundle* of views paired with four-dimensionalism, the very same *bundle* paired with three-dimensionalism is metaphysically equivalent: any version of three-dimensionalism is equivalent to an analogous version of four-dimensionalism and *vice versa*.

In chapter one I argued that correctly inter-translatable theories are metaphysically equivalent. The problem lies in determining whether some function that maps sentences from one theory onto sentences of some other theory, is a correct translation. To that end I introduced the notion of both an assertibility mapping and a practical translation. An assertibility mapping, recall, is a function that maps the sentences of one theory onto the sentences of another theory just when those sentences are assertible under the same possible situations. A practical translation is an assertibility mapping that is truth preserving. So all correct translations are practical translations, but not all practical translations are correct translations, since a practical translation may preserve truth in virtue of the existence of different truth makers. The task of arguing that we can infer that there exists a correct translation from the existence of a practical translation is one that will be undertaken in chapter six. In this chapter I aim only to show that there exists a practical translation between analogous three- and four-dimensionalist theories. To that end I appeal to the various diagnostic criteria discussed in chapter one.

In this chapter I focus on showing that there exists a practical translation between analogous *unitary* three- and four-dimensionalist theories. This involves further refining our account of different unitary theories. In chapter four I developed an account of unitary three-dimensionalism in both restricted and unrestricted versions. But I provided the merest sketch of a four-dimensionalist account. This chapter develops such a view. This view is largely a novel one.[1] And it is, I think, particularly interesting on two counts.

One reason to be interested in the view that persisting objects are four-dimensional, yet lack temporal parts, is that such a view appears to provide a plausible middle-ground between traditional three-dimensionalism and traditional four-dimensionalism. In virtue of this, *prima facie* it seems that it might be able successfully to preserve a greater portion of our intuitions about objects and persistence. For instance, the unitary four-dimensionalist can agree with the traditional three-dimensionalist that an ontology of temporal parts is both counterintuitive and metaphysically profligate. She can agree that where we see a single persisting object, there really *does* exist a single persisting object and not a myriad of temporally overlapping objects constantly passing into and out of existence. She can also concur with the three-dimensionalist in denying that persisting objects are only partly present whenever they exist. On the other hand, denying that persisting objects are only partly present whenever they exist, does not entail affirming that they are wholly present whenever they exist. So the unitary four-dimensionalist is not committed to defending, or indeed defining, the notion of being wholly present. Moreover, since the unitary four-dimensionalist denies that persisting objects are strictly identical to themselves at every time at which they exist, she is not faced with any problem of reconciling strict identity across time with change. Yet she can agree with the three-dimensionalist that when we attribute some simple intrinsic property to an object, we are attributing it to the persisting object itself, not to some part of that object.

So one good reason to consider versions of unitary four-dimensionalist is that at least prima facie, there are some reasons to suppose that such theories might be plausible accounts of persistence. In fact, although I think that unitary four-dimensionalism is a coherent view, it is not a very compelling view. For the same sorts of reasons that I prefer non-unitary three-dimensionalism to unitary three-dimensionalism, I prefer non-unitary four-dimensionalism to unitary four-dimensionalism. And if this is right then it tells us something too: we should be non-unitary theorists.

Before I turn to develop these unitary theories, some programmatic issues. First, within the class of unitary theories there are a number of further distinctions to be drawn. In chapter four we saw that unitary theories come in both restricted and unrestricted varieties. Thus already we have two unitary three-dimensionalist theories, and two unitary four-dimensionalist theories. Then each of these theories comes in a presentist and an eternalist version. This would bring the number of unitary three-dimensionalist theories to four, and similarly for unitary four-dimensionalist theories. In general, although I explicitly define and discuss restricted and unrestricted versions of unitary three- and four-dimensionalism, I do not spend the same amount of time defining presentist and eternalist versions. I do not, for instance, define and name any theory that is a presentist, unrestricted, unitary three-dimensionalism. This is in part because I think that the issue of presentism versus eternalism is relatively clear cut with respect to theories of persistence: given that the key definitions we constructed in chapter three are consistent with presentism and eternalism, it is relatively straightforward to see, for each different version of three- or four-dimensionalism, what a presentist or eternalist version of that theory would look like.

So although I do consider some of the issues that arise from adopting different theories of time, I spend less time on this distinction than on some others. As a default I tend to consider versions of unitary theories that adopt eternalism rather than presentism, except where considering the latter is especially pertinent. This is because at least *prima facie*, eternalism seems to present obstacles for the three-dimensionalist in a way that it does not for the four-dimensionalist, whereas presentism seems equally consistent (odd, but consistent) with both three- and four-dimensionalism. Thus the case *against* three- and four-dimensionalism being metaphysically equivalent is, as I see it, stronger given an eternalist reading of each, and so this is where I focus much of my attention.

Second, one of the diagnostic criteria for a practical translation is empirical equivalence. If theories are practically inter-translatable, this entails that they are (strongly) empirically equivalent—they make the same observational predictions in all possible worlds. So if it turns out that any analogous versions of three- and four-dimensionalism are not empirically equivalent, then this entails that these versions are not practically, and thus not correctly, inter-translatable, and hence not metaphysically equivalent.

It is certainly difficult to see how any empirical data could reveal whether the filled region of space before me contains an *O*-stage, or a wholly present object *O*. Intuitively, at least, it seems that there is no possible observational prediction that would be a prediction of three- but not four-dimensionalism (in any version) or *vice versa*. So it is tempting to conclude that the two theories are empirically equivalent, and thus that this criterion is fulfilled. And in this chapter I will assume just that: that this necessary but insufficient condition for practical inter-translatability and hence metaphysical equivalence obtains.

It has been argued, however, that various empirical discoveries such as those that led confirm the theory of special relativity, are either inconsistent with three-dimensionalism, or at the least strongly militate in favour of four-dimensionalism. Showing that this is not so requires a whole chapter, and I leave this task until chapter eight. For now I assume that our intuitions are correct, and that analogous versions of three and four-dimensionalism are empirically equivalent.

Let us turn then, to develop unitary versions of three- and four-dimensionalism.

1. REFINING UNITARY THREE-DIMENSIONALISM

Recall that unitary three-dimensionalism, is roughly the view that there exist at least some fusions of temporally overlapping enduring simples, and that there exist at least some enduring non-fusions that are constituted whenever they exist, by those fusions at those times. Let us return to reconsider world w in which there exist three enduring simples A, B and C. Recall that A endures from t_0 to t_5, B endures from t_1 to t_6, and C endures from t_2 to t_7. And recall that O that exists between t_2 and t_5. and has spatial parts A and B at t_2, B and C at t_3, A and C at t_4, and A, B and C at t_5. In chapter four, I argued that the unitary three-dimensionalist cannot say that O is the fusion of A B and C, (call that fusion *ABC*) since that fusion exists from t_0 to t_7 and has parts A B and c at t_4. This is what led us to develop the idea that O is constituted by different enduring fusions at each of the times at which it exists, (it is constituted by fusion *AB* at t_2, by *BC* at t_3 by *AC* at t_4 and *ABC* at t_5).

But how does all of this fit in with the apparatus we constructed in chapter three? In chapter three I drew a distinction between having parts M-*simpliciter* and having them S-*simpliciter*. When we consider enduring fusions such as *ABC*, we can see exactly how this apparatus comes into play. Consider first the notion of having parts M-*simpliciter*. It is tenselessly true of enduring fusion *ABC*, that *A*, *B* and C, are parts of that object—that is what it is to be a fusion of *A*, *B* and C. This is the sense in which *A*, *B* and *C* are parts of *ABC* M-*simpliciter*. Even at t_7 and t_0 when *B* does not exist, it is still true at those times, that *B* is part of *ABC*: *B* is part of *ABC* M-*simpliciter*. There is also a sense in which *B* is not part of *ABC* at t_0 and t_7, and this is the sense in which *B* is not part of *ABC* S-*simpliciter* at those times. For *B* is not had $t_0$ly or $t_7$ly, and thus is not had S-*simpliciter* at t_0 or t_7.

Thus a fusion has parts M-*simpliciter* and S-*simpliciter* just if:

> Fusion M-*Simpliciter*: A fusion *F* has a part *P* M-*simpliciter* just if *F* fuses *P*.
>
> Fusion S-*Simpliciter*: A fusion *F* has a part *P* S-*simpliciter* at a time *t* just if *F* has *P* M-*simpliciter*, and *F* has *P* *t*ly at *t*.

What of enduring non-fusions such as O? We know that non-fusions have parts in virtue of being constituted by fusions that have parts. What does this tell us about the parts that non-fusions like *O* have M- and S-*simpliciter*? Well in the case of non-fusions, the notion of parthood M-*simpliciter* is less useful, since by definition non-fusions are not fusions of particulars, it is not tenselessly true that they have the parts that they do. So is there any sense in which *O* has, say, part *A* M-*simpliciter*?

As we noted in chapter three, it is certainly true that *O* has *A* $t_2$ly (and $t_4$ly and $t_5$ly). Doesn't that mean that *O* has *A* M-*simpliciter*? Yes and no. The only candidates for being tenseless claims about *A* and *O*, are the claims that *O* has *A* at t_2, or that *O* has *A* at some time. So to say that *O* has *A* M-*simpliciter* would really be to say no more than that *O* has *A* at t_2 M-*simpliciter*. But though this is true, it is not really the sense of M-*simpliciter* that we are looking for. We might conclude, therefore, that when it comes to non-fusions, the M-*simpliciter* notion fails to track anything useful. In a sense this is true. It is unprofitable to talk about the parts that non-fusions have M-*simpliciter*. It is worth noting though, that non-fusions inherit from the fusions that constitute them at times, the idea of simply having a part, rather than having a part relative to some time. Consider *O* at t_2 once more. At t_2*O* is constituted by the fusion *AB*. *AB* has parts *A* and *B* M-*simpliciter*: it has them in a straightforward non-temporally modified manner. So at t_2, although we do not want to say that *O* has parts *A* and *B* M-*simpliciter*, we can say that *O* is constituted at t_2 by some fusion that has parts *A* and *B* M-*simpliciter*. What explains why *O* does not have part *A* at t_2, but rather, just straightforwardly has *A* at t_2—has *A*$t_2$ly—is that at t_2 there is some object, *AB*, that has *A* M-*simpliciter*. It is the fact that non-fusions 'borrow' from the fusions that constitute them at times, this metaphysically basic sense of having a part, that explains why enduring non-fusions have parts, at times (have them *t*ly) rather than having parts-at-times.

This leaves us with explicating the notion of having a part S-*simpliciter* within the context of enduring non-fusions. This is straightforward. Roughly, what it is for a non-fusion to have a part S-*simpliciter* at a time is for the fusion that constitutes that object at that time to have the part M-*simpliciter*, and for the fusion to have that part *t*ly at *t* (to have that part S-*simpliciter*) This brings us back to the sense we discussed in chapter four, in which we get to talk about non-fusions having parts at times in virtue of being constituted by fusions at times—the sense in which they 'inherit' those parts. We can now explicate more thoroughly this idea of 'inherited parts'. For we can say the following:

> Non-fusion S-*simpliciter*: An enduring non-fusion *O* has a part *P* S-*simpliciter* at time *t* just if (i) at *t O* is constituted by some fusion *F* that has *P* M-*simpliciter* and (ii) at *t F* has *P* S-*simpliciter*.

Hence we get to say that *O* has part *A* at t_2 (it has *A* S-*simpliciter* at t_2), and it does so in virtue of being constituted at t_2 by fusion *AB* which has *A* S-*simpliciter* at t_2. This is the sense in which *O* 'inherits' part *A* at t_2, from fusion *AB*.

So let us now move on to develop a unitary version of four-dimensionalism.

2. UNITARY FOUR-DIMENSIONALISM

I noted in chapter four that unitary versions of four-dimensionalism hold that persisting objects are temporally extended temporal simples: four-dimensional objects that have spatial parts but lack temporal parts.[2] Unitary four-dimensionalism, then, is the view that objects persist by *terduring*. In chapter three we defined terdurance as follows:

> T: An object *O* terdures just if it exists at multiple temporal locations in virtue of being a temporally extended temporal simple that does not wholly exist at any of the temporal locations at which it exists.

Of what are terduring objects composed? Well the unitary four-dimensionalist rejects the claim that for any persisting object *O*, there exist instantaneous objects that overlap *O* whenever it exists: they reject the non-unitary thesis. So the unitary four-dimensionalist rejects the idea that there exist any fusions-at-times of persisting objects. What does this tell us about the relation between terduring objects and the objects of which they are composed? Indeed, what are terduring objects?

Well, since terdurantists deny that there exist any instantaneous objects that overlap persisting objects at times, we know that if they hold that composite objects are composed entirely of simples, then these simples must themselves persist. This follows from the argument we considered in chapter four. On the other hand, if composite objects are composed entirely of gunk rather than simples, then we know that any four-dimensional volume of gunk is not diachronically divisible: we know that the object is composed of four-dimensional gunk streams. This must be so if the terdurantist is to avoid commitment to the existence of maximal temporal parts.

So let us assume that our terdurantist is an atomist. Then she holds that simples persist. How do they persist? Well, if we are terdurantists, then they must persist by terduring. What is it to terdure? Clearly:

A simple *S* terdures just if:

(i) *S* is mereologically simple and
(ii) *S* is four-dimensional.

Then surely some composite terduring objects are fusions of terduring simples. But then consider a world w^* that looks like world w. w^* contains only three persisting simples—*A*, *B*, and *C*—but unlike w, w^**'s* simples terdure. Just as in w, *A* persists from t_0 to t_5, *B* persists from t_1 to t_6, and *C* persists from t_2 to t_7. Putting aside issues of under what circumstances composition occurs, there exist various fusions of these terduring simples, fusions *AB*, *BC*, *AC*, and *ABC*. These terduring objects—terduring fusions, as we might call them—are objects that have as spatial parts at a time, those simples that exist at that time and are fused by that object. So, for instance, terduring fusion *ABC* has no (maximal) temporal parts, but it does have spatial parts: namely at t_2, it has spatial parts A, *B* and C. Now suppose we posit the existence of another terduring object, call it *O*. *O* looks like *O* in w, insofar as it exists between t_2 and t_5 and has spatial parts *A* and *B* at t_2, *B* and *C* at t_3, *A* and *C* at t_4, and A, *B* and *C* at t_5. But *O* in w^* terdures.

This leaves the unitary four-dimensionalist with an analogous problem to that faced by unitary three-dimensionalists, namely what to say about objects such as *O* that are not fusions of enduring simples (in the case of the three-dimensionalist) or terduring simples (in the case of the four-dimensionalist). And just as unitary three-dimensionalism appeals to some relation that holds between enduring fusions and non-fusions at times, so too the unitary four-dimensionalist will need to appeal to some relation that holds between terduring fusions and terduring non-fusions at times.

So what is the relation between terduring objects *O* and *ABC*, and between *O* and *AB*? *O* is neither identical to fusion *ABC* nor to *AB*. Nor do *O* and

AB overlap in the sense of sharing a (maximal) temporal part at t_2. Roughly, the relation between *O* and *AB* is that at t_2, *O* and *AB* share all of the same spatial parts. Once again, though, the four-dimensionalist is faced with a problem analogous to that faced by the three-dimensionalist. In chapter four we asked the question: how can it be that non-fusions have parts at times, given that what it is for particular *P* to be a part of *O*, is for *P* to be fused by *O*? And we answered this question by appeal to the constitution relation. The same question arises here. How can it be that *O* has spatial parts *A* and *B* at t_2, given that *O* is not a fusion of *A* and *B*? The answer must reside in whatever relation it is that holds between *O* and *AB* at t_2. Call this relation the *compilation relation*. Then plausibly:

> Compilation: A fusion *F* compiles a (terduring) non-fusion *O* at a time *t* just if (i) *F* and *O* exist at *t*, and (ii) the existence of *F* entails the existence of *O* at *t* and (iii) there is no proper part of *F* whose existence entails the existence of *O* at *t*.

Just as for the three-dimensionalist, what explains how enduring non-fusion *O* has parts at times is that *O* is constituted by some fusion at those times, so too for the four-dimensionalist, what explains how terduring non-fusion *O* has parts at times is that *O* is compiled by some fusion at those times. And just as the constitution relation appeals to the fact that what it is for some enduring non-fusion to exist at a time is for there to exist some fusion at that time, so too the compilation relation appeals to the idea that what it is for some terduring non-fusion to exist at a time is for there to exist some fusion at that time. The truth maker for the existence of some terduring non-fusion at some time *t*, is that there exists some (terduring) fusion at *t* which entails the existence of the non-fusion at that time.

Moreover, just as the three-dimensionalist requires some analysis of the relation that holds between enduring non-fusions at times (such as between an enduring lump and statue) so too the four-dimensionalist will require a similar analysis. For most four-dimensional persisting objects will turn out to be terduring non-fusions: objects that strongly gain and lose parts. If the statue and lump are four-dimensional objects, then they are likely terduring non-fusions. The relation between terduring non-fusions then, will appeal to the idea of compilation. For the unitary four-dimensionalist will say that two or more terduring non-fusions that materially coincide at a time, do so in virtue of being compiled by the same fusion at that time. That is, what it is for two or more terduring non-fusions to materially coincide at a time, is to be related in the same way to some further object at that time. Call this relation the *co-compilation relation*. Hence:

Co-compilation: For any two or more terduring non-fusions O and O^* that exist at a time t, O and O^* are co-compiled at t just if (i) O is compiled by some fusion F at t, and (ii) O^* is compiled by some fusion F^* at t, and (iii) F is identical to F^*.

We will return to this notion of co-compilation shortly.

So notice a couple of things here. Like the unitary three-dimensionalist, the unitary four-dimensionalist is committed to holding a particular view about simples, or about gunk: she is committed to holding that simples persist, or to holding that four-dimensional hunks of gunk are divisible in only certain ways: it is not diachronically divisible. So she is committed to a particular view about the way the world is at the fundamental level. She is also committed to holding that there exist multiple kinds of persisting objects: terduring fusions and terduring non-fusions. Further, these objects are related in a certain way: terduring non-fusions are compiled at times by terduring fusions. This means commitment to the existence of the compilation relation. So the unitary four-dimensionalist's picture of the world is substantially different to that of the traditional four-dimensionalist. It is a picture according to which there exist ontologically basic, massively scattered fusions of terduring simples, which, at certain times, entail the existence of certain non-fusions that are the ordinary objects of our experience. What is interesting about this view is that once we actually spell out the details of a unitary four-dimensionalism, it turns out to involve rather more than just the claim that persisting objects are four-dimensional, yet lack (maximal) temporal parts: it commits us to a whole range of other claims. Those who find this view unpalatable might then conclude that it is not simply that traditional four-dimensionalism is preferable, but perhaps more strongly, that temporal parts are an integral and deeply important part of the four-dimensionalist theory. Once we attempt to develop a four-dimensionalist theory that does not include the apparatus of temporal parts, we find we have a radically different theory, which, it might be argued, is unappealing in various ways.

Now that we have a general idea of what unitary four-dimensionalism looks like, we can define two versions of the theory. These definitions will be important in the following sections where we begin to show how the translation between unitary versions of three- and four-dimensionalism will proceed.

In fact, we are defining what we might think of as 'pseudo' restricted and unrestricted versions of unitary four-dimensionalism. These versions are pseudo unrestricted insofar as both place some restriction on composition: neither allow that there can be diachronic fusions of temporally non-overlapping objects, and hence that there can be any perduring objects. Thus:

RCU4D: (i) for some sets $S \ldots S_n$ of temporally overlapping terduring simples, there exist the fusions $F \ldots F_n$ of the members of those sets and (ii) for some fusions $F \ldots F_n$ and times $t \ldots t_n$ at which those fusions exist, there exist terduring non-fusions $O \ldots O_n$ which are compiled by those fusions at those times.

Restricted unitary four-dimensionalism is the view that at least some terduring simples can be fused to create terduring fusions, and at least some terduring fusions compile terduring non-fusions at times. Unrestricted complex unitary four-dimensionalism is the view that:

UCU4D (i) for any arbitrary set S of temporally overlapping terduring simples, there exists a fusion of the members of S and (ii) for any arbitrary set S of fusions, and any arbitrary times at which those fusions exist, there exists some terduring non-fusion that is compiled by those fusions at those times.

Unrestricted unitary four-dimensionalism is the view that all temporally overlapping terduring simples can be fused to create terduring fusions, and that whenever such fusions exist, they compile some terduring non-fusion at that time.

With these definitions in mind, it is now possible to see whether we have a practical translation between the various unitary three-dimensionalist views and their four-dimensionalist analogues. In the next section I begin this task by providing an assertibility mapping between analogous theories. Once we have this mapping we can determine whether the mapping meets the diagnostic criteria of a practical translation: whether the mapping gives us theories that are equally explanatory, and whether it preserves the principles of charity and humanity.

In determining whether the theories meet the first of these diagnostic criteria—explanatory equivalence—I consider how the explanatory apparatus of each unitary theory deals with the four puzzle cases we have met. The hope is that in considering the explanatory apparatus of each of these theories, we will not only be able to determine whether they are explanatorily equivalent, but in addition, we will get a much more detailed picture of each theory as an account of persistence. We will see just how each theory accounts for various phenomena of persistence: how each makes sense of cases of fission and how each understands the relation between objects that coincide at times. This will allow to determine whether unitary theories are plausible accounts of persistence, and to begin to answer the question of whether they are preferable to non-unitary accounts. So investigating the issue of whether analogous unitary theories are equivalent or not, allows us in the process to examine some novel accounts of persistence.

3. DIAGNOSING A PRACTICAL TRANSLATION

3.1 Assertibility Mappings

Looking at the definitions in the previous section, we can see just how similar the different versions of unitary three- and four-dimensionalism really are.[3]

Notice first that the unitary three- and four-dimensionalist who embraces unrestricted composition will agree about which persisting objects exist: each will assert under all and only the same conditions, that there exists some persisting fusion, and each will assert under all and only the same conditions, that there exists some persisting non-fusion. The same is true for unitary three- and four-dimensionalists who embrace the same form of restricted composition. (For the definitions of restricted unitary three- and four-dimensionalism are consistent with a range of different restrictions on composition.) Three- and four-dimensionalists who agree on how composition is restricted, will also agree about which persisting fusions and non-fusions exist. So any restricted unitary three-dimensionalist who holds that composition is restricted in way *W*, will assert that there exists some persisting fusion just when the restricted unitary four-dimensionalist who holds that composition is restricted in way *W*, will assert that there exists some persisting fusion, and so too *mutatis mutandis* for persisting non-fusions.

In essence, the only difference between the unitary three-dimensionalist and the unitary four-dimensionalist is that when the former asserts that there exists a persisting fusion, she asserts that there exists some *enduring* fusion, while when the latter asserts that there exists a persisting fusion, she asserts that there exists some *terduring* fusion. Thus the three-dimensionalist will assert that there exists an enduring fusion just when the four-dimensionalist will assert that there exists a terduring fusion, and the three-dimensionalist will assert that there exists an enduring non-fusion, just when the four-dimensionalist will assert that there exists a terduring non-fusion. Moreover, since the *only* difference between the compilation and the constitution relation is that the former holds between terduring fusions and terduring non-fusions at times, and the latter holds between enduring fusions and enduring non-fusions at times (that is, the only difference between the relations is their relata) it follows that the three-dimensionalist will assert that some persisting non-fusion is constituted by some fusion at a time, just where the four-dimensionalist will assert that some persisting non-fusion is compiled by some fusion at a time.

So, for instance, consider world w_1 (which is just like w and w^* except we do not specify that the simples either endure or terdure), which contains three temporally overlapping persisting simples *A*, *B*, and *C*. Then the unrestricted unitary three-dimensionalist will hold that in w_1 there exist four fusions *AB*, *BC*, *AC*, and *ABC*. The unrestricted unitary four-dimensionalist will agree that these are all and only the fusions that exist in w_1. In addition to these fusions, both (unrestricted) unitary three- and four-dimensionalists will hold that there exist a number of non-fusions. They agree that there exists an object that is related at one time, to fusion *AB*, at another time to fusion *BC*, and at another time, to fusion *AC*. Indeed, they agree about exactly which non-fusions exist. The only thing they truly disagree about is whether the fusions and non-fusions endure or terdure (they disagree about how these objects are related: by constitution or compilation, but the only difference between these two relations is their relata, so ultimately it comes down to a difference between terdurance and endurance).

So there is an assertibility mapping between the notions of compilation and constitution, and thus also between the notions of co-compilation and co-constitution. There is also an assertibility mapping between the notions of a terduring fusion and an enduring fusion, and a terduring non-fusion and an enduring non-fusion. Given this, it is easy to see how the definitions of restricted unitary three-dimensionalism and restricted unitary four-dimensionalism map onto each other, and how the definitions of unrestricted unitary three-dimensionalism and unrestricted unitary four-dimensionalism map onto each other.

Indeed, the only difference between these definitions is that one appeals to terduring fusions that compile terduring non-fusions at times, while the other appeals to enduring fusions that constitute enduring non-fusions at times. Since the only difference between the compilation and the constitution relation is in terms of their relata, it follows that, unsurprisingly, the only real difference between these two views is that one holds that persisting objects endure, and the other that persisting objects terdure. Ultimately then, it all comes down to the difference between terdurance and endurance: if unitary three- and four-dimensionalism are distinct, it is because there is some genuine difference between enduring and terduring as a manner of persistence. If the views are equivalent then it is because talk of objects enduring and terduring is really just two ways of talking about the same manner of persistence.

So what is the difference between enduring and terduring? We know that endurance is defined in terms of objects being wholly present at a time, while terdurance is defined in part in terms of objects failing to wholly exist at a time. *Prima facie*, it seems plausible that failing to wholly exist at a time, is the same as failing to be wholly present at a time, in which case objects endure only if they do not terdure and *vice versa*. Of course, in chapter three I left undefined the four-dimensionalist's notion of failing to wholly exist at a time. So let us put aside that notion for a moment and return to the three-dimensionalist's notion of endurance.

In chapter three I defined endurance in terms of the notion of being wholly present at a time, which is in turn defined in terms of an object having all of its parts S-*simpliciter* present at a time. What is the unitary four-dimensionalist to make of the notion of having a part S-*simpliciter*? Notice that just as the unitary three-dimensionalist wants to be able to talk about the spatial parts, at a time, of an enduring fusion, so too the four-dimensionalist wants to be able to talk of the spatial parts at a time, of a terduring fusion. But terduring fusions, like enduring fusions, have their parts tenselessly: it is tenselessly true of terduring fusion *ABC* that it has *A*, *B* and *C* as parts, even though there are times at which only *A* exists (eg t_0). The sense in which *A*, *B* and *C* are tenselessly parts of *ABC* is different to the sense in which at t_0, *A* is part of *ABC* and *B* and *C* are not. Earlier in this chapter I explicated this difference within the context of unitary three-dimensionalism, in terms of persisting fusions having parts M- and S-*simpliciter*. That is, I defined the following notions:

> Fusion M-*Simpliciter*: A fusion *F* has a part *P* M-*simpliciter* just if *F* fuses *P*.
>
> Fusion S-*Simpliciter*: A fusion *F* has a part *P* S-*simpliciter* at a time *t* just if *F* has *P* M-*simpliciter*, and *F* has *P* *t*ly at *t*.

It appears that the unitary four-dimensionalist will want to avail herself of the same distinction. That is, the unitary four-dimensionalist will want to say that the fusion *ABC* tenselessly has parts *A*, *B* and *C*, (and thus has them at t_0) and this sense is the sense in which it has these parts M-*simpliciter*. She will also want to say that at t_0 there is a sense in which *ABC* has part *A*, and lacks parts *B* and *C*. This is the sense in which at t_0 *ABC* has part *A* S-*simpliciter*, and lacks parts *B* and *C* S-*simpliciter*. Both unitary three- and four-dimensionalists can embrace the definitions above, the only difference being that the four-dimensionalist will hold that the fusion is a terduring fusion, while the three-dimensionalist will hold that the fusion is an enduring fusion.

What of non-fusions and their spatial parts at times? In the case of perduring objects we can think of the spatial parts of an object at a time as being parts *simpliciter* of the relevant temporal part of that object (or as being non-maximal temporal parts *simpliciter* of the perduring object). This is the sense in which the perdurantist gets to say that objects straightforwardly have parts, rather than having them relative to some time. For the unitary four-dimensionalist, however, it clearly cannot be that terduring non-fusions that are compiled at times by terduring fusions, have their spatial parts *simpliciter*, for non-fusions do not tenselessly have parts in the way that fusions do. Yet we want to be able to talk about the spatial parts of non-fusions at times, where those parts are had in a perfectly straightforward, non-relativised manner. The four-dimensionalist does not want to talk of non-fusions having parts-at-times, but rather, of having parts *simpliciter*, at times. Hence she must talk of having parts *simpliciter* in different temporally modified ways—of having parts *t*ly. The unitary four-dimensionalist must embrace some sort of adverbialism about the parts of non-fusions. The sense in which non-fusions have parts *(simpliciter) t*ly, is the sense in which they have parts S-*simpliciter* at times. Earlier in this chapter I defined the notion of enduring non-fusions having parts S-*simpliciter*. The unitary four-dimensionalist will surely adopt a similar account according to which:

> Non-fusion S-*simpliciter*: A terduring non-fusion *O* has a part *P* S-*simpliciter* at a time *t* just if (i) at *t O* is compiled by some fusion *F* that has *P* M-*simpliciter* and (ii) at *t F* has *P* S-*simpliciter*.

So the four-dimensionalist gets to say that persisting non-fusions have parts *simpliciter* at times—they straightforwardly have those parts at times by having them *t*ly at *t*, in virtue of being related at those times, to some fusion that has those parts both M- and S-*simpliciter*.

What then, is the relationship between enduring and terduring? We know that enduring objects are wholly present whenever they exist, in virtue of having all of their parts S-*simpliciter* present whenever they exist. Do terduring objects have all of their parts S-*simpliciter* present whenever they exist? They do: what it is for a terduring object to exist at a time, is to have all of its parts S-*simpliciter* at that time. Like all fusions (including enduring ones), terduring fusions exist at a time just if at least one of theirs parts M-*simpliciter* exists at that time. Fusion *ABC* exists at t_0 because *A* exists at t_0. That is, *ABC* exists at t_0 because *ABC* has part *A* S-*simpliciter* at t_0, and it fails to exist at t_{11} because it has no parts S-*simpliciter* at t_{11}. Moreover, *ABC* has all of its parts S-*simpliciter* at t_0, just as it does at every other time. The same is true for non-fusions. Indeed, for *any* persisting non-fusion *O*, what it is for *O* to exist at a time is for *O* to have all of its parts S-*simpliciter* present at that time. *O* has all of its parts S-*simpliciter* at a time just if it is related at that time, to some fusion that has those parts M- and S-*simpliciter* at that time, and where that fusion *entails* the existence of the non-fusion at that time. Hence what it is for a terduring object to exist at a time, is the same as what it is for an enduring object to exist at a time. So if, in the following sections, I successfully argue that analogous versions of unitary three- and four-dimensionalism are practically inter-translatable, what does this tell us about the relation between being wholly present at a time, and failing to wholly exist at a time?

If the mapping between the notions of endurance and terdurance is truth preserving, then these cannot be contrary notions. Rather, we might suppose that what it is for a terduring object *O* to fail to wholly exist at a time *t*, is for *O* tenselessly to have parts—to have parts M-*simpliciter*—and thus for it to be true at *t*, that *O* is composed of parts some of which do not exist at *t*. Terduring fusions, then, fail to wholly exist at a time. Then perhaps a terduring non-fusion *O** fails to wholly exist at a time *t*, if *O** is compiled at *t*, by a fusion that does not wholly exist at *t*. But of course, the endurantist does not disagree with any of *that*: if that is what it is to fail to wholly exist at a time, then she can agree that enduring objects do not wholly exist at a time even though they are wholly present whenever they exist.

Ultimately, since composite objects are composed of simples, the distinction between endurance and terdurance comes down to the distinction between an enduring simple and a terduring simple. Four-dimensionalists think simples terdure: that they are temporally extended and do not wholly exist whenever they exist. Three-dimensionalists think that simples endure, and are wholly present whenever they exist. The problem is that it is difficult to see just what the crucial difference is supposed to be. The four-dimensionalist talks of simples being temporally extended, but if all *that* means is that they exist at multiple times (since it cannot mean that they partly exist at those times in virtue of some part existing at those times) then the three-dimensionalist will agree with that—after all, that is what she means when she says that simples endure. And what can it mean to say that simples are wholly present, or fail to wholly exist when they exist? What distinction could that capture, given that both views concede that simples exist at multiple times and lack both spatial and temporal parts? It is difficult to see in virtue of what there could be any real distinction here. So if we find good reason to suppose, given the criteria of practical inter-translatability, that a practical translation exists, then we have good reason to suppose there is no difference.

So here we have a plausible assertibility mapping. The first question to address is whether that mapping preserves truth—is it a practical translation—and if it is a practical translation, is it also a correct translation? In order to answer these questions we must consider whether these theories meet the other diagnostic criteria of practical inter-translatability that I set out in chapter one. As I noted at the beginning of this chapter, I set aside the criterion of empirical equivalence until chapter eight. In this chapter I concentrate on the other criteria of a practical translation, focusing a good deal of attention on the criterion of explanatory equivalence. While

I argue that there exists a practical translation between analogous versions of unitary three- and four-dimensionalism, I do not yet attempt to show that this practical translation is a correct one. I leave that task to chapter six, where I consider the broader issue of what licenses the inference from the existence of a practical translation to the existence of a correct translation.

I turn now to consider whether unitary three- and four-dimensionalism are equally explanatory, specifically, whether they have the same explanatory resources in dealing with each of the puzzles that I described in chapter two. As we noted in chapter one, we can rule out as practically inter-translatable, theories (which are on the same level) for which there is an assertibility mapping but where those theories have different degrees of explanatory power. Thus discovering that co-assertible theories are equally explanatory provides some reason to suppose that they are practically translatable and perhaps ultimately metaphysically equivalent.

3.2 Explanatory Equivalence

Illustrating that two theories are explanatory equivalent is no easy task: it is not possible to consider every possible explanandum, and then show that the theories provide equally explanatory resources for dealing with that explanandum. To limit this task, I will consider the four key puzzles of persistence that we discussed in chapter two. If it can be shown that unitary versions of three- and four-dimensionalism have the same explanatory resources with respect to these key four problems, then this is good evidence that they are equally explanatory. So let us turn first to the problem of temporary intrinsics.

3.2.1 Temporary Intrinsics

How do unitary versions of three- and four-dimensionalism deal with the problem of temporary intrinsics? In chapters two and three I described the manner in which 'traditional' three-dimensionalism approaches the problem of temporary intrinsics, namely either by relativising properties to times—indexicalism—or relativising the having of those properties to times—adverbialism. Unitary three-dimensionalism will approach this problem in basically the same way. Or, I should say, eternalist versions of unitary three-dimensionalism will approach the problem in the same way, since, arguably, presentist versions are not faced with the same difficulty. After all, the argument goes, if only the present exists then the only properties that an object has, are the ones that it has in the present, and there is therefore no concern about reconciling change over time with strict identity over time.

(In fact I remain puzzled by this response. An object that exists in the present counts as being a *persisting* object only if either it did exist in the past, or will exist in the future. But then it is still the case that an object that exists in the present, is identical to an object that did exist in the past and will exist in the future. And the object that did exist in the past and will exist in the future, had different properties when it existed in the past, from the properties that it has in the present, and that object will have different properties when it exists in the future, from the properties it has in the present. So at different times at which a persisting object exists, it has different properties at some of those times.[4] Thus if the presentist can really make sense of the idea that objects persist, then the problem of temporary intrinsics ought to be just as much of a problem for him as for the eternalist: it is the problem of how an object that did exist in the past with one set of properties, can be identical to an object that does exist in the present, and has a different set of properties. Of course, nothing I say hangs on this being the case. If presentists are not faced with the problem of temporary intrinsics, then there is no worry for the three-dimensionalism presentist and hence the problem of temporary intrinsics is no problem at all. If presentists are faced with this problem, then they can deal with it in the same way as the eternalist. A way that I shall now move on to discuss.)

We know that the 'traditional' three-dimensionalist resolves the problem of temporary intrinsics by appeal to adverbialised (or temporally relativised) properties that are instantiated at every time at which an enduring object exists. I will talk only of adverbialism, since that view has the advantage that it allows that persisting objects instantiate properties *simpliciter*, they merely instantiate them in different temporally modified ways. Recall that in chapter three I argued that three-dimensionalists should distinguish between instantiating properties M-*simpliciter* and instantiating them S-*simpliciter*. I will not rehearse this account at length here. What is clear is that the unitary three-dimensionalist will appeal to this same apparatus. What though, of the unitary four-dimensionalist? To answer this question we need to start by thinking about spatially extended mereological simples.

Suppose there exist spatially extended simples, and suppose those simples have different intrinsic properties at different spatial locations. Let us say that there exists a simple, *MS*, that is red at one location and green at another. We cannot analyse these local intrinsic properties in terms of the existence of some spatial part at one location that is red *simpliciter*, and a spatial part at another location that is green *simpliciter*. Rather, we must appeal to some sort of manifestation or instantiation relation. Plausibly, we might appeal to the idea of such simples instantiating properties at a spatial location, (at-*S*) or instantiating them in a particular spatial-locational manner (*S*ly). That is, we might appeal to spatial analogues of indexicalism or adverbialism. Thus we might say that *MS* has the property of redness at-S_1 or S_1ly, (where S_1 is the location at which the simple is red) and has the property of greenness at-S_2 or S_2ly (where S_2 is the location at which the simple is green).

The same will be true of persisting simples. To explain and analyse *temporary* intrinsics, unitary four-dimensionalists must appeal to some sort of instantiation relation. Just as spatially extended simples have no spatial parts that instantiate properties *simpliciter*, so temporal simples have no temporal parts that instantiate properties *simpliciter*. So if intrinsic properties are not to turn out to be disguised relations to times, then the instantiation relation had better be adverbial. Hence the unitary four-dimensionalist should hold that properties are instantiated in particular temporal ways: namely $t_1$ly, $t_2$ly $t_3$ly and so forth. What it is for a terduring object to be green at one temporal location and red at another, is for that object to be green in one temporally modified way, and red in a different temporally modified way.

In appealing to an adverbialist analysis of property instantiation, the four-dimensionalist may be tempted to adopt the strategy I have recommended for the three-dimensionalist, according to which the notion of *simpliciter* is not univocal. If we suppose that at t_1 some terduring object O^* is red, and at t_2 is green, then O^* is red $t_1$ly and green $t_2$ly—it is red *simpliciter* and green *simpliciter*, but in different temporally modified ways. So there is a sense in which it is straightforwardly the case that O^* is both red and green. But the sense in which it is tenselessly true that O^* is red and green, is different to the sense in which at t_1, we would judge that O^* is red and is not green. Hence the four-dimensionalist might say that the sense in which if it is tenselessly the case that O^* is red $t_1$ly then it is red *simpliciter*, is the sense in which O^* is red M-*simpliciter*. Thus the sense in which O^* is both red and green, is the sense in which O^* is red M-*simpliciter* and green M-*simpliciter*. But the sense in which we would judge, at t_1, that O^* is red and not green, is the sense in which O^* is red S-*simpliciter* at t_1, but is not green S-*simpliciter* at t_1—it is the sense in which O^* is red $t_1$ly at t_1, and is green $t_2$ly at t_1.

It looks then, as though both unitary three- and four-dimensionalist deal with the problem of temporary intrinsics in the same way, by appeal to an irreducible notion of having properties in particular temporally modified ways—having them *t*ly. Notice that despite the fact that persisting non-fusions (whether enduring or terduring) have properties at times in virtue of being related in certain ways (compiled or constituted) to persisting fusions at those times, nevertheless non-fusions still instantiating properties in an irreducibly temporally adverbialised manner. Why so? Well persisting non-fusions are constituted or compiled by persisting fusions which themselves have irreducibly adverbialised properties. Persisting fusions, whether terduring or enduring, will likely have different properties at different times, and hence the properties of these fusions will themselves be temporally adverbialised. So while the properties of non-fusions are reducible to the properties of the fusions that constitute or compile them at times, this does not mean that the properties of non-fusions are not *irreducibly* adverbialised. Both unitary three- and four-dimensionalists will appeal to irreducibly adverbialised properties.

The point is that unitary three- and four-dimensionalists will resolve the problem of temporary intrinsics in the same manner: they have the same explanatory resources with respect to this puzzle. So let us now turn to consider the puzzles of temporary and permanent coincidence.

3.2.2 Temporary and Permanent Coincidence

This brings us to the problems of temporary and permanent coincidence. Consider first temporary coincidence. In general the problem of temporary coincidence is couched in terms of the coincidence of ordinary persisting objects that gain and lose parts across time: objects such as a statue and a lump, a person and a body and so forth. As I noted previously, the sorts of objects that we find in these examples are almost certainly non-fusions—for they strongly gain and lose parts. But let us return to these non-fusions in a moment. First, let us begin by considering the temporary coincidence of persisting fusions.

Consider the persisting fusions *AB* and *ABC* (ignoring for the moment whether they endure or terdure). At t_1 *A* and *B* exist but *C* does not. So at t_1 fusion *AB* and fusion *ABC* are materially coincident. Recall that in chapter one I noted that for the 'traditional' four-dimensionalist, cases of both temporary and permanent coincidence are understood in terms of temporary and permanent overlap, where persisting objects temporarily overlap just if there are times at which they share (maximal) temporal parts, and permanently overlap just if there are no times at which they fail to share (maximal) temporal parts. The unitary three- and four-dimensionalist can say something similar, albeit without reference to temporal parts. They too can understand temporary and permanent coincidence in terms of overlap, as each can say that at t_1 *ABC* and *AB* overlap. In fact, at t_1 *ABC* and *AB* do not just overlap—they do not just share a part at t_1—rather, they *completely* overlap at t_1: they share all of the same parts. Since objects completely overlap at a time just if they are materially coincident at that time, we might think that talking of overlap adds nothing to the explanatory mix. In a sense that is true. In practical terms though, talk of objects coinciding at times is frequently associated with the view that such objects are related at those times, by some 'mysterious' relation such as constitution. Though I do not see that constitution is a mysterious relation, I note here that even if it were, the complete overlap of fusions at times makes no reference to such a relation: it is a straightforward case of overlap.

Since *AB* and *ABC* are fusions of different particulars, they must be distinct objects. Indeed, even at t_1 when they coincide and may *appear* to be identical, it is still the case that *AB* and *ABC* have different parts M-*simpliciter*, thus explaining why they are distinct. At t_1 what explains why *AB* and *ABC* share the same intrinsic properties is that they completely overlap at that time—they share the same parts S-*simpliciter* at that time. So both unitary three- and four-dimensionalists will hold that the temporary coincidence of persisting fusions can be understood in terms of the complete overlap at times, of those fusions: their analyses are analogous.

What of the permanent coincidence of persisting fusions? I will return to re-address this question when I consider an analogous question about the permanent coincidence of non-fusions. For now, it seems that cases of permanent coincidence simply cannot occur within the context of persisting fusions. For any two fusions F_1 and F_2 are identical just if they are fusions of all and only the same particulars—if they share the same parts M-*simpliciter*. So it follows that there cannot be any cases where *distinct* persisting fusions permanently coincide, since what it is for fusions to permanently coincide is for them to share all of the same parts M-*simpliciter*. So it seems that both unitary three- and four-dimensionalists will reject the idea that distinct persisting fusions can permanently coincide.

Usually though, talk about temporary or permanent coincidence is talk about the coincidence, at times, of persisting non-fusions such as the lump and the statue. What does the unitary three-dimensionalist make of cases where enduring non-fusions temporarily coincide? Unlike in the case of enduring (and terduring) fusions, the relation between enduring non-fusions that coincide at times cannot be explicated in terms of overlap. The three-dimensionalist must appeal to some further relation. This relation is not, however, the constitution relation. Non-fusions are constituted at times, by fusions. But non-fusions do not constitute other non-fusions at times. Rather, non-fusions coincide at times in virtue of being constituted at those times, by one and the same fusion. Thus the relation that holds between coincident enduring non-fusions at times is the co-constitution relation, a relation I defined earlier:

> Co-constitution: For any two or more enduring non-fusions O and O^* that exist at a time t, O and O^* are co-constituted at t just if (i) O is constituted by some fusion F at t, and (ii) O^* is constituted by some fusion F^* at t, and (iii) F is identical to F^*.

Thus when the statue and the lump coincide they are co-constituted in virtue of each being related in the same way—by constitution—to the same fusion at that time. The statue and lump share the same intrinsic properties when they coincide because at that time they are constituted by the same fusion, and thus each has the intrinsic properties of that fusion at that time—that is, each has all of the properties S-*simpliciter* that the fusion has S-*simpliciter* at that time. Yet the statue and the lump are distinct, since there are times when they are constituted by distinct fusions: times when they are not co-constituted. It is in virtue of the fact that there are times at which they are not co-constituted, that even at the time of coincidence the statue and the lump have different properties. For enduring non-fusions have irreducibly temporally adverbialised properties that they instantiate at every time at which they exist. The statue and the lump have *different* adverbialised properties in virtue of there being times at which they are not co-constituted, and thus they have different properties even when they are co-constituted.

The unitary four-dimensionalist will explicate temporary coincidence in an analogous manner. But rather than appealing to the co-constitution relation, she will appeal to the co-compilation relation that I defined earlier in this chapter:

> Co-compilation: For any two or more terduring non-fusions O and O^* that exist at a time t, O and O^* are co-compiled at t just if (i) O is compiled by some fusion F at t, and (ii) O^* is compiled by some fusion F^* at t, and (iii) F is identical to F^*.

Thus the four-dimensionalist will hold that terduring non-fusions are co-compiled at a time just if at that time each is related to—compiled by—one and the same fusion. As with the three-dimensionalist then, coincident non-fusions share the same intrinsic properties at a time in virtue of each being compiled by the same fusion at that time. So too terduring non-fusions that temporarily coincide are nevertheless distinct, since there are times at which they are not co-compiled. Since terduring non-fusions have irreducibly temporally adverbialised properties, terduring non-fusions that only temporarily coincide will have different adverbialised properties even at the times at which they coincide.

So far we have shown that unitary three- and four-dimensionalism have the same explanatory resources for dealing with the temporary coincidence of both persisting fusions and non-fusions. But what of the permanent coincidence of non-fusions, and what, if anything, is the relation between the permanent coincidence of fusions, and the permanent coincidence of non-fusions?

A typical example of permanent coincidence is the case of Lumpl and Goliath which are, likely, persisting non-fusions. What ought the unitary three- and four-dimensionalist make of such cases? There are three options. The first is to say that in cases of permanent *non-fusion* coincidence we really have a case of identity. On this view, a non-fusion O_1 is identical to a non-fusion O_2, just if O_1 and O_2 exist at all and only the same times, and at each of those times are constituted by, or compiled by, the same fusions. This option involves treating persisting non-fusions in the same way as persisting fusions, and thus concluding that distinct persisting objects *never* permanently coincide. The task for the proponent of this view is to explain away the different modal properties of the statue and the lump as mere appearance, or to argue that such modal properties are not 'real' properties of objects, but rather, are merely the result of our conventions.[5]

The second option is to hold that Lumpl and Goliath are distinct, since they do instantiate genuinely different modal properties. What explains why they appear to be identical is that they are related by co-constitution or co-compilation at every time at which they exist. Thus they share all of the same intrinsic properties *simpliciter*, the only properties they do not share are modal properties, and modal properties are not observable. On this view, there is something substantially different about persisting fusions and persisting non-fusions. Distinct persisting fusions never permanently coincide, while distinct persisting non-fusions, such as Lumpl and Goliath, may permanently coincide. The question is whether this is a plausible distinction, and if so, what it is that undergirds that distinction. I will return to this issue shortly.

Finally, the third option is to hold that Lumpl and Goliath are contingently identical. Contingent identity is, in broad strokes, the thesis that a single individual may have two different designations in the actual world, which, under different circumstances, may have referred to two distinct individuals. The contingent identity thesis is the thesis that *statements* of identity can be contingent. Identity itself is necessary. It will always be necessary that x is identical to x, that is, that x is self identical. What is contingent is that the designation 'y' picks out x in the actual world. So what might have been two distinct individuals is, in the actual world, one individual whose dual designations are contingently identical.

There are a number of different ways of explicating the thesis of contingent identity, some of which do not appeal to the existence of other possible worlds.[67] By far the clearest account, however, is Lewis', and it is this account that I will consider. According to Lewis, names have associated with them various counterpart relations: that is, a name picks out in other possible worlds, individuals relevantly similar to the one picked out in the actual world, and these counterparts represent what is possible for the individual in the actual world.[8] In the case of Goliath/Lumpl we have two names attached to one individual and each of these names has associated with it a particular counterpart relation—each picks out different counterparts in different possible worlds. Depending on whether we are interested in Goliath/Lumpl qua statue or qua lump, we will be interested in a different counterpart, for in many possible worlds that individual will have two counterparts, a lump counterpart and a statue counterpart.

The idea is that the designations 'Lumpl' and 'Goliath' pick out the same non-fusion in the actual world, but there are worlds in which they pick out distinct objects—lump and statue counterparts. What is contingent is that the designation 'Lumpl' picks out same individual in the actual world, as the designation 'Goliath'. Recourse to contingent identity thus dissolves the Lumpl and Goliath puzzle by acknowledging on the one hand that Lumpl and Goliath are identical—there really does exist only one individual—Goliath/Lumpl—in the actual world—and yet on the other hand that had things been different, there would have been two distinct individuals that could rightly have been referred to as 'Goliath' and 'Lumpl.'

Which of these options ought the unitary three- and four-dimensionalist accept? For our purposes it doesn't really matter. So long as each of these options has the same costs and benefits for the unitary three-dimensionalist as it does for the unitary four-dimensionalist, all that matters is that each has the same options available, and thus that each has the same explanatory resources when it comes to the puzzles of temporary and permanent coincidence.

But which *should* they choose? Consider first the view that Lumpl and Goliath are distinct. We could make a case for treating persisting fusions and non-fusions differently in the face of permanent coincidence. The idea would be that since fusions are nothing over and above the particulars they fuse, it makes no sense to think that two distinct fusions could exist, but nevertheless be composed of all and only the same particulars. On the other hand, non-fusions are such that they could have been compiled or constituted at times, by different fusions from the ones they are in fact related to. Thus two persisting non-fusions might be permanently coincident, but they are nevertheless distinct because in different worlds they are not co-compiled or co-constituted. This is something like what the contingent identity theorist holds, except that she thinks that in the actual world the permanently coinciding non-fusions are identical.

The difficulty, however, with holding that permanently coinciding non-fusions are distinct becomes apparent when we consider unrestricted versions of unitary three- or four-dimensionalism (and indeed many restricted versions of each). Unrestricted unitary theories hold that there exist fusions of any arbitrary temporally overlapping persisting simples and that those fusions always constitute or compile some persisting non-fusion at a time. Thus there will exist many odd, spatially and temporally scattered persisting non-fusions that are related at times, to fusions. But is there any reason to suppose that there could exist two or more such gerrymandered non-fusions that permanently coincide? If we admit that Lumpl and Goliath are distinct, then we should also admit that there exist many permanently coinciding non-fusions. Indeed, we should think that for every persisting non-fusion *NF*, there exist a number of distinct non-fusions that permanently coincide with *NF* in the actual world (though do not permanently coincide with *NF* in other worlds).

After all, ontology ought to be independent of human convention, so it cannot be that Lumpl and Goliath are distinct objects purely in virtue of the existence of some *conventions* regarding lumps and statues, ie. that statues do not survive being squashed. Our conventions may render that particular modal property salient, but absent the convention it would still be the case that some actual non-fusion ceases to exist in a counter-factual world in virtue of being constituted or compiled in that world, by a fusion that has the property of being flat. We do not confer modal properties on objects, our conventions merely make salient, properties that already exist. Just because we are not familiar with the modal properties of gerrymandered objects does not mean that they do not have such properties. But then it looks like ontological profligacy wins the day. For not only is it the case that there exist multiple permanently coinciding gerrymandered non-fusions, (presumably as many distinct objects as there are worlds in which those objects have different properties) but it is also the case that where there exists both Lumpl and Goliath permanently coinciding, so too there exist other permanently coinciding objects with other modal properties, merely properties that are less salient to us that those associated with lumps and statues. I take it, though, that such profligacy is unattractive.

It strikes me that the contingent identity thesis has the advantage of allowing that these modal properties are genuine properties, while also allowing us to staunch the flow of ontological profligacy. We get to say that all of these modal properties are real, in that these non-fusions have different counterparts in different worlds. Some of these counterparts are salient to us, as in the Lumpl and Goliath case where we care about lump counterparts and statue counterparts. But all of them exist, even counterparts of gerrymandered objects. But this does not mean that in the actual world these objects are distinct and that there exist a plethora of distinct permanently coinciding objects that differ only with respect to these modal properties. Rather, in the actual world these objects are identical.

Of course, what matters is that the explanatory resources of unitary three- and four-dimensionalism are equivalent with respect to these particular puzzles, and they are.

I turn then, to consider our final puzzle: fission.

3.2.3 Fission

Fission, recall, is the process whereby an object effectively 'splits' into two qualitatively identical objects each of which is related to the pre-fission object in the same manner, and is qualitatively identical to the pre-fission object just prior to fission. Hence in chapter two we discussed this in relation to the circumstance in where Will Riker undergoes transportation and two qualitatively identical persons—R_1 and R_2—each of whom is psychologically continuous with pre-fission Riker, result from the process. I noted that it is claimed to be a virtue of four-dimensionalism, or more correctly perdurantism, that it can analyse what occurs in cases of fission in a way that preserves all our folk intuitions on the matter, and a vice of three-dimensionalism that it cannot. But what of *unitary* theories' analyses of fission?

Consider first persisting fusions. It is not clear that strictly speaking, fusions can undergo fission. But something that looks like fission might occur. Suppose that there exist six persisting simples *A, B, C, D, E* and *F*, where *A* and *B* exist from t_1 to t_5, and *C, D, E* and *F* exist from t_6 to t_{10}. Then if we suppose that *A* and *B*, *C* and *D*, and *E* and *F* are all arranged in the same manner, (the fusions of *CD* and *EF* are qualitatively identical to

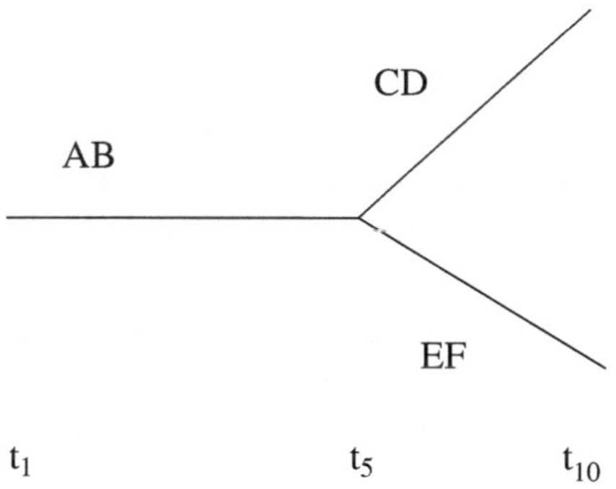

the fusion of *AB*) and that *C*, *D*, *E* and *F* come into existence at (roughly) the same spatial location at which *A* and *B* passed from existence, then we have a situation in which it appears that *AB* splits into two qualitatively identical objects, *CD* and *EF*.

But there is nothing mysterious occurring here. It is just a case in which there exist two fusions—*ABCD* and *ABEF*—each of which temporarily overlap between t_1 and t_5. (We could also describe this as a case in which fusion *ABCDEF* becomes spatially scattered at t_6 in a way that it was not prior to that time, and described in this way we do not have a case of fission at all—one man's fission is another man's spatial scattering.) In fact, though, neither the unitary three- nor four-dimensionalist will describe the case above in the manner just described. For both unitary three- and four-dimensionalists disavow the existence of fusions of temporally non-overlapping objects. Thus both admit that there exist fusions *AB*, *CD* and *EF*, but not fusions *ABCD* or *ABEF*: for these would be fusions of temporally non-overlapping simples, and thus objects that perdure. So unitary theorists of both stripes will say that this is straightforwardly a case in which fusion *AB* ceases to exist at one time, and two new fusions, *CD* and *EF* come into existence at a temporally contiguous time.

Indeed, when it comes to persisting fusions, unitary three- and four-dimensionalists will hold that it is never the case that a fusion can survive fission. For if *C*, *D*, *E* and *F* temporally overlap with *A* and *B* then although fusions *ABCD* and *ABEF* exist, this will not count as a case of fission. For the 'post-fission' objects *CD* and *EF* would not be qualitatively identical with the 'pre-fission' object, since that object would, prior to t_6, have parts *A, B, C, D, E* and *F*.

When we are considering cases of fission though, we are usually considering the fission of non-fusions such as Riker. What are unitary three- and four-dimensionalists to make of such cases? Since fission is just a particular instance of temporary overlap (one in which each of the post-fission objects is related in the same manner to the pre-fission object), both unitary views will treat the Riker case in an analogous manner. For the unitary four-dimensionalist cannot appeal to the apparatus of temporal parts to analyse fission. Rather he will describe this as a case in which prior to fission, both R_1 and R_2 are co-compiled, and after fission R_1 and R_2 are not co-compiled, but rather, each is compiled after fission by a different fusion. Thus both R_1 and R_2 exist prior to fission, and are identical with R_1 and R_2 respectively after fission. This view has basically the same virtues as the traditional perdurantist account insofar as it allows that Riker survives fission: 'Riker' refers ambiguously to R_1 and R_2, and both R_1 and R_2 survive fission. So too the unitary three-dimensionalist will describe this case in an analogous manner, except that she will say that R_1 and R_2 exist prior to fission and are at those times co-constituted, while after fission R_1 and R_2 are no longer co-constituted.

What of my objection in chapter one, that since for the three-dimensionalist there must be some fact of the matter prior to fission as to how many objects exist and are constituted (or co-constituted) at those times, then given the possibility of fission there must always exist multiple coinciding objects? Is this a problem for the unitary three-dimensionalist in a way that it is not for the unitary four-dimensionalist, and is there any solution to the problem? I will not answer these questions here. Since the same issue arises for non-unitary theories, I will consider this matter in detail in the following chapter. Let us provisionally accept that in chapter six I will resolve this issue satisfactorily, and thus provisionally accept that unitary three- and four-dimensionalism resolve cases of fission in an analogous manner.

So far we can say that with respect to all of our four puzzle cases, analogous versions of three- and four-dimensionalism are equally explanatory. Plausibly then, unitary three- and four-dimensionalism are equally explanatory *simpliciter*. This, in turn, gives us reason to suppose that our assertibility mapping amounts to a practical translation. There is, however, one final diagnostic criterion of a practical translation: appeal to the principle of charity. It is to that I now turn.

3.3 The Principle of Charity

The principle of charity tells us that we should expect most of the sentences that others assert to be true, and therefore that we should interpret them thus. The more plausible principle of humanity tells us that we should expect most of the sentences that others assert not to be inexplicably false, and that we should interpret them thus. That is, unless we have reason to suppose that the other party is suffering under hallucination, or is factually mistaken, or has made a mistake of logic or metaphysics, then we should apply to her utterances the principle of charity.

Despite the fact that complex unitary three and four-dimensionalism are co-assertible theories, and are (we have so far assumed) empirically equivalent and are, as I have just argued, equally explanatory, it might still be argued that the theories are not practically inter-translatable because while one theory is true, the other is false. If that were the case, then by the lights of the four-dimensionalist, proponents of three-dimensionalism would frequently utter falsehoods, and by the lights of the three-dimensionalist, proponents of four-dimensionalism would frequently utter falsehoods.

But is such a scenario plausible? Well, are there any facts that once made clear to the three-dimensionalist, would result in her recognising that some of her utterances were, by her own lights, false? It is difficult to imagine that there are. The three-dimensionalist is not empirically mistaken: she is not suffering from hallucination or illusion, she does not mistakenly believe that science reveals that objects endure; nor is she making a mistake of logic or metaphysics: she does not misunderstand Leibniz' Law, or fail to grasp the axioms of mereology. And the same surely goes for the four-dimensionalist. The point is that it is impossible to imagine any piece of information that one party might convey to the other, such that the second party comes to realise that they are mistaken in their view of persistence. (If that were the case, then surely four-dimensionalists would long ago have converted three-dimensionalists or *vice versa*).

Given this, the principle of humanity tells us that we ought to interpret the utterances of both three- and four-dimensionalist as true.[9] This provides additional reason to suppose that the assertibility mapping is truth preserving, and thus that we have a practical translation. This should hardly come as a shock. The sole difference between terduring and enduring objects is that terduring objects do not wholly exist whenever they exist, while enduring objects are wholly present whenever they exist: both enduring and terduring objects are persisting temporal simples that have spatial parts at times. It is difficult to see what this supposed difference between being wholly present and not wholly existent at a time could amount to.

Ultimately though, the question still remains as to whether this practical translation is a correct one: are unitary versions of three- and four-dimensionalism really metaphysically equivalent? I return to this question—the crux of the matter—in chapter six after I have first developed non-unitary versions of three- and four-dimensionalism, and then considered whether these versions are equivalent.

NOTES

[1] Though Parsons (2000) argues that there could be a version of four-dimensionalism that rejects the existence of temporal parts, ultimately the view he describes is rather different to the view I develop.

[2] More precisely, they lack maximal temporal parts: temporal parts that overlap every spatial part at a time, of the persisting object of which they are a part. For spatial parts are, of course, non-maximal temporal parts of four-dimensional objects.

[3] We only need to compare some of the definitions to see this. Compare the definitions of unrestricted unitary three-dimensionalism and unrestricted unitary four-dimensionalism:

UU4D (i) for any arbitrary set *S* of temporally overlapping terduring simples, there exists a fusion of the members of *S* and (ii) for any arbitrary set *S* of fusions, and any arbitrary times at which those fusions exist, there exists some terduring non-fusion that is compiled by those fusions at those times.

UU3D (i) for any arbitrary set *S* of temporally overlapping[3] simples, there exists a fusion of the members of *S* and (ii) for any arbitrary set *S* of fusions, and any arbitrary times at which those fusions exist, there exists some enduring non-fusion that is constituted by those fusions at those times.

We find the same analogous structure for the definitions of constitution and compilation.

[4] It has been suggested to me (by Ted Sider) that this is the step that the presentist rejects, since she rejects the idea that 'it was the case that x is F' needs to be cashed out in terms of x having F *at* a moment of past time. Rather,'it was the case that x is F' stands in need of no further analysis. But then it seems you ought to be able to say both that it was the case that x is F, and it was the case that x is not F (so long as there was a time at which 'x is F' is true, and another at which 'x is not F' is true. Maybe there is no contradiction in holding both that 'it was the case that x is F' and 'it was the case that x is not F'. Maybe this would only be a contradiction if one said that 'it was the case that x is F and x is not F.' So, perhaps presentists don't have a problem here. I think I just can't take tense seriously enough.

[5] See for instance Heller (1990) chapter one, and (1984).

[6] Gibbard (1975); Noonan (1991).

[7] Though I remain unclear how one is ultimately to make sense of these other accounts—Abelardian predicates, for instance—without appeal to some sort of counterpart theory.

[8] Lewis (1986).

[9] Or at least, their utterances about the world, not their meta-utterances to the effect, for instance, that they disagree with one another. (Those utterances are surely false if they interpret the utterances about the world as true).

Chapter 6

THE METAPHYSICAL EQUIVALENCE OF NON-UNITARY THREE- AND FOUR-DIMENSIONALISM

1. NON-UNITARY VIEWS

It is now time to turn our attention to non-unitary versions of three- and four-dimensionalism. Non-unitary theories hold that at every time at which persisting objects exist, they do so in virtue of being related to some instantaneous object—a synchronic fusion—that exists at that time.

So what does a non-unitary view of persistence entail with respect to other metaphysical commitments? Well first, the issue of whether a commitment to a non-unitary theory entails anything about the nature of simples, depends on how one construes the debate between fusions-at-times and fusions at times. Suppose that in fact simples persist. Then do there exist any instantaneous objects that are synchronic fusions of those simples at a time? The unitary theorist says no: she denies that there can exist any fusions-at-a-time of objects that exist at other times. That is, she doesn't simply deny that one needs to be committed to such objects, she denies that one can be committed to such objects: for it simply makes no sense to talk of a fusion-at-a-time: there can only ever exist fusions *simpliciter*. If one accepts this reasoning, then one must conclude that given atomism, non-unitary theories entail that simples are instantaneous. This does not entail that simples are point-sized. It is consistent with this view that simples might be spatially extended, but not temporally extended.

On the other hand, one might reject the unitarist's reasoning, and hold that there can exist fusions-at-a-time: there can exist synchronic fusions of persisting simples. Then any view about simples is consistent with a non-unitary theory. Any view about *simples* is consistent with a non-unitary theory, but this does not mean that there are not additional commitments that the non-unitarist must embrace. For if she thinks that there can exist fusions-at-times, then she must think that there are some additional mereological axioms which allow us to create fusions-at-a-time.

Similarly, if one holds that composite objects are composed entirely of gunk, then one must hold not only that gunk is diachronically divisible, but that it is divisible at every instant. That is, there exist instantaneous objects composed of gunk, and these objects are related to persisting objects in certain ways.

Some non-unitarists are also committed to making or rejecting certain claims about composition. Three-dimensionalists, for instance, are committed to rejecting the idea that composition across time is mereological in nature: instead, she holds that composition is non-mereological. Four-dimensionalists, on the other hand, are committed to composition across time being mereological.

So we can see that embracing a non-unitary view entails embracing certain other views. What I argued in chapter four, however, was that it does not preclude one from embracing either restricted or unrestricted composition. Thus in chapter four we defined unrestricted endurantism as the view that embraces non-mereological universalism (that is, non-mereological universalism just is unrestricted endurantism by another name) and unrestricted perdurantism as the view that embraces mereological universalism (that is, unrestricted perdurantism just is mereological universalism by another name). Thus unrestricted perdurantism is the view that:

> UP: for any arbitrary time t and set S of concrete particulars there exists a fusion of the members of S at t (a synchronic fusion-at-a-time) and for any arbitrary set S^* of synchronic fusions that exist at distinct times $t_1 \ldots t_n$, there exists a diachronic fusion of the members of S^*.

Unrestricted endurantism is the view that:

> UE: for any arbitrary time t and set S of concrete particulars there exists a fusion of the members of S at t (a synchronic fusion-at-a-time) and for any arbitrary set S^* of synchronic fusions that exist at distinct times $t_1 \ldots t_n$, there exists an enduring object O that is at each of those times, constituted by one of those fusions.

Hence both unrestricted perdurantist and unrestricted endurantist hold that any arrangement of particulars that exists at a time, composes some instantaneous object, and any arrangement of instantaneous objects across time is related in some way, to some persisting object. For the unrestricted perdurantist these instantaneous objects are temporal parts of persisting objects, for the endurantist they constitute persisting objects at times.

Then restricted perdurantism is the view that:

> RP: for some times $t \ldots t_n$ and some sets $S \ldots S_n$ of concrete particulars, there exist synchronic fusions $F \ldots F_n$ of the members of those sets at those times and (ii) for some synchronic fusions $F \ldots F_n$ that exist at distinct times $t_1 \ldots t_n$, there exist diachronic fusions of those fusions.

And restricted endurantism is the view that:

> RE: for some times $t \ldots t_n$ and some sets $S \ldots S_n$ of concrete particulars, there exist synchronic fusions $F \ldots F_n$ of the members of those sets at those times and (ii) for some synchronic fusions $F \ldots F_n$ that exist at distinct times $t_1 \ldots t_n$, there exist enduring objects $O \ldots O_n$ that are constituted at each of those times, by those fusions.

These definitions tell us that at least some arrangements of particulars at times compose instantaneous objects, and that at least some of these instantaneous objects are related in some way to persisting objects. (Notice that just as in the case of restricted unitary three- and four-dimensionalism, these definitions leave it open that for *every* synchronic fusion that exists, there exists at that time some persisting object that has that synchronic fusion as a temporal part, or which is constituted at that time by that fusion.)

With these definitions in place, it is possible to develop assertibility mappings between restricted endurantism and restricted perdurantism, and between unrestricted endurantism and unrestricted perdurantism. We can then move on to determine whether those mappings are truth preserving.

Once again, in determining whether the mappings preserve truth, we will consider at length whether analogous non-unitary theories are equally explanatory. In doing so, we will attain a much better picture of non-unitary theories as accounts of persistence. We will take a detailed look at the explanatory apparatus of both three- and four-dimensionalist non-unitary theories, and ultimately this will put us in a position to determine whether either of these theories is plausible, and if non-unitary theories are more plausible than unitary theories.

1.1 Assertibility Mappings

According to the non-unitary theorist, in virtue of what do persisting objects exist? For the endurantist, persisting objects exist at times in virtue of being constituted by some synchronic fusion at those times. For the perdurantist, persisting objects exist at times in virtue of being composed of some temporal part at those times. An enduring object is constituted by a (synchronic) fusion at some time t, if that fusion entails the existence of the enduring object at that time. We could make a similar claim with respect to perduring objects: a perduring object has a temporal part at some time t, if there exists some synchronic fusion that entails the existence of the perduring object at that time. *Prima facie* then, it seems that we have an assertibility mapping between the three-dimensionalists' notion of constitution, and the four-dimensionalists' notion of temporal parthood. That is, whenever the endurantist will assert that some persisting object is constituted by some synchronic fusion at some time t, the perdurantist will assert that some persisting object has some synchronic fusion as a temporal part at t.

It does appear, however, that there will be problems in translating restricted and unrestricted versions of endurantism into restricted and unrestricted versions of perdurantism. The unrestricted endurantist will assert that there exists some persisting object just when the unrestricted perdurantist will make the same assertion. Both agree that there exist a myriad of odd gerrymandered objects, some of which are synchronically gerrymandered (like the trout-turkey) and some of which are diachronically gerrymandered (they have temporal parts, or are constituted at times, by synchronic fusions that are not causally related in the usual manner of everyday objects). So too restricted versions of each (which place the same restriction on composition), will agree about which synchronic fusions and persisting objects exist.

The only difference is that the endurantist holds that persisting objects are *constituted* by synchronic fusions at times, while the perdurantist holds that persisting objects have synchronic fusions as *temporal parts* at times. This difference, however, marks a crucial difference between the mereological axioms that each view accepts. The endurantist rejects certain mereological axioms: namely, she denies that it is possible to fuse instantaneous objects that exist at different times—she denies that there exist any diachronic fusions. All endurantists (restricted and unrestricted alike) reject mereological universalism, embracing instead a non-mereological account of composition across time. All perdurantists embrace a mereological conception of composition across time, and hence affirm the existence of diachronic fusions. Given that these views accept different mereological axioms, how could it be that they are, as I claim, correctly inter-translatable?

At the beginning of the chapter I defined unrestricted perdurantism. Notice though, that we could have defined unrestricted perdurantism as follows:

> UP(a): for any arbitrary time t and set S of concrete particulars, there exists a fusion of the members of S at t (a synchronic fusion-at-a-time) and for any arbitrary set S^* of synchronic fusions that exist at distinct times $t_1 \ldots t_n$, there exists a perduring object O that at each of those times has as a temporal part one of those synchronic fusions.

Though equivalent to UP, this amended definition is analogous to the definition of unrestricted endurantism. Unsurprisingly, comparing UP(a) with UE reveals that the crucial issue is whether there is a mapping between the three dimensionalist's notion of an enduring object, and the four-dimensionalist's notion of a perduring object.

Let us begin then, defining parthood M- and S-*simpliciter* in the context of non-unitary three-dimensionalism. In fact, with respect to fusions, these two notions are the same for the non-unitary as for the unitary three-dimensionalist. Hence:

> Fusion M-*Simpliciter*: A fusion *F* has a part *P* M-*simpliciter* just if *F* fuses *P*.
>
> Fusion S-*Simpliciter*: A fusion *F* has a part *P* S-*simpliciter* at a time *t* just if *F* has *P* M-*simpliciter*, and *F* has *P* *t*ly at *t*.

What of enduring objects? Enduring objects have parts at times, in virtue of being related at those times, to fusions that have those parts (M)-*simpliciter*. In chapter five I defined the notion of parthood S-*simpliciter* within the context of unitary three-dimensionalism. We can appeal to the same definition:

> Enduring S-*simpliciter*: An enduring non-fusion *O* has a part *P* S-*simpliciter* at time *t* just if (i) at *t O* is constituted by some fusion *F* that has *P* M-*simpliciter* and (ii) at *t F* has *P* S-*simpliciter*.

(Notice that clause (ii) is unnecessary in this context. Unlike unitary three-dimensionalists, non-unitary three-dimensionalists hold that since all fusions are instantaneous, they have the same parts M-*simpliciter* as they do S-*simpliciter*[1])

There is no straightforward sense in which enduring non-fusion *O* itself has parts M-*simpliciter* (in the same way that there was no straightforward sense, for the unitary three-dimensionalist, in which persisting non-fusions have parts M-*simpliciter*). But it is tenselessly true that *O* is constituted by some fusion *F* at some time *t*, and that at *t P* is part of *F* M-*simpliciter*. That is, for instance, at t_1 it is true of *O* that at t_2 *O* is constituted by some fusion that has *P* M-*simpliciter*. At present this might not seem a perspicuous use of the locution 'M-*simpliciter*', but let us use the terminology since this notion will become important. Then we can say that:

> Non-unitary Enduring M-*Simpliciter*: A non-unitary enduring object *O* has a (spatial) part *P* M-*simpliciter* just if there is some time *t* at which *O* is constituted by a fusion *F*, and *F* has P M-*simpliciter*.

Now let us consider perdurantism. For the perdurantist too, there is a tenseless sense in which perduring objects have spatial parts at times. This is the sense in which *P* is a (spatial) part of perduring object *O* just if *P* is part of some temporal part of *O*—*P* is a non-maximal temporal part of *O*. But this is not the only sense of 'having a part'. The perdurantist will want to distinguish between the sense in which *O* has *P* as a part when it has *some* temporal part that has *P* as a part, and the sense in which *O* has *P* as a part when it has some *current* temporal part that has *P* as a part. To distinguish these two senses, let us call the former having a part M-*simpliciter*, and the latter having a part S-*simpliciter*. Then:

> Perduring M-*simpliciter*: A perduring object *O* has a (spatial) part *P* M-*simpliciter* just if there is some time *t* at which *O* has as a temporal part synchronic fusion *F*, and *F* has *P* M-*simpliciter*.

Then the sense in which *P* is part of *O* at *t* just if *P* is part of the temporal part of O that exists at *t*, is captured by the S-*simpliciter* sense of having a part:

> Perduring S-*simpliciter*: A perduring object *O* has (spatial) part *P* S-*simpliciter* at *t* just if *O* has as a temporal part some synchronic fusion *F* that exists at *t*, and *F* has *P* M-*simpliciter*.

Consider the following scenario. Suppose there exists some persisting object *O* that has some (proper) spatial part *P* at t_1, and lacks *P* at t_2. Endurantists hold that *O* is an enduring object, and typically hold that at t_1, the claim '*P* is part of *O*' is true, and that at t_2 the claim '*P* is part of *O*' is false. Perdurantists hold that *O* is a perduring object, and typically claim that '*P* is part of *O*' is tenselessly true, and thus true at both t_1 and t_2. (For notice that *P* is a non-maximal temporal part of *O*, and hence it is tenselessly true that *P* is part of *O*.) But if perdurantists have a tenseless view of parthood, and endurantists a tensed view, how could the two views be equivalent?

The answer is clear: endurantists and perdurantists mean something different by 'part'. Or, given the distinctions we have just drawn, endurantists and perdurantists disagree about which notion of parthood captures the everyday sense of the term. Endurantists think that '*P* is part of *O*' is false at t_2, because they interpret 'part' in this context in terms of parthood S-*simpliciter*: they think the everyday sense of 'part' is captured by the S-*simpliciter* sense. Thus they think that parthood is tensed. Perdurantists think that '*P* is part of *O*' is true at t_2, because they think that 'part' in this context should be interpreted as part M-*simpliciter*, and at t_2, *P is* part of *O* M-*simpliciter*. Perdurantists think that the everyday sense of 'part' is captured by the M-*simpliciter* sense of parthood, and thus that parthood is tenseless.

Once we see this, we can see that there is nothing contradictory in the way that endurantists and perdurantists understand spatial parthood: for both of their claims are true under the correct interpretation. Indeed, we can see how an assertibility mapping will proceed. The endurantists will say that *P* is part of *O* S-*simpliciter* just when the perdurantists will say that *P* is part of *O* S-*simpliciter* and *mutatis mutandis* for parthood M-*simpliciter*. Once we see that endurantist and perdurantist simply mean something different by 'part' we see that they are not making contradictory claims at all.

What of temporal parts? Suppose that *P* is a synchronic fusion that exists at t_1. For the perdurantist, it is tenselessly true that *P* is (a temporal) part of *O*: for *O* fuses *P*. So for the perdurantist, the natural reading of '*P* is part of *O*' is such that the claim is true at t_1, the time at which *P* exists, and also true at t_2, the time at which *P* does not exist. The endurantist, however, does not think that synchronic fusion *P* is part of *O*. Or at least, she does not think that *P* is a proper part of *O*. Rather, she thinks that *P* constitutes *O* at t_1 and thus that *P* is an improper part of *O* at t_1. So the endurantist thinks that the natural reading of '*P* is part of *O*, is such that the claim is true at t_1, but false at other times.

It looks as though there is genuine disagreement between endurantist and perdurantist. Notice though, that even the perdurantist will admit that there is *some* sense in which we might want to affirm, at $t_{1,}$ that *P* is part of *O*, and deny, at t_2, that *P* is part of *O*. It is natural to draw this distinction in terms of having parts M- and S-*simpliciter*. For the perdurantist, the 'natural' reading of '*P* is part of *O*' is captured by the M-*simpliciter* sense: the sense in which it is tenselessly true that *P* is part of *O*. The other sense is best captured by the S-*simpliciter* sense: the sense in which '*P* is part of *O*' is true at t_1 just if *P* exists at t_1 and *P* is part of *O* M-*simpliciter*. This is the sense in which *P* is part of *O* at t_1, and is not part of *O* at t_2.

Prima facie, it looks as though the endurantist's notion of having a part is similar to the perdurantist's notion of having a part S-*simpliciter*. But can the endurantist draw a similar distinction to the one drawn by the perdurantist? Well the endurantist can certainly say that it is tenselessly true that at t_1, *P* constitutes *O*—that is, it is tenselessly true that at t_1, *P* is an improper part of *O*. Thus she can say that *P* is part of *O* M-*simpliciter*, where *P* is part of *O* M-*simpliciter* just if there is some time *t* at which *P* is part of *O*. This notion of M-*simpliciter*, however, does not seem to capture the same sense as the perdurantist's notion. After all, the endurantist only gets to talk about the parts M-*simpliciter* of enduring object *O*, if she includes a temporal index in the definition of parthood M-*simpliciter*. Ultimately, to say that *P* is part of *O* M-*simpliciter* is really just to say that there is some time *t* at which *P* is part of *O*. The perdurantist however, requires no temporal index in the notion of having a part M-*simpliciter*: for her, *P* is part of *O* M-*simpliciter* just if *O* fuses *P*—that is, if *P* is a temporal part of *O*.

So while the endurantist can only make the tenseless claim that 'at t_1, *P* is part of *O*', the perdurantist can make the more straightforward tenseless claim that '*P* is part of *O*'. Notice, however, that the claim that '*P* is part of *O*' is actually the claim that *P* is an instantaneous *temporal* part of *O*. What is it to be an instantaneous (maximal) temporal part? It is to be an object that exists at, and only at a certain instant, and which overlaps at that instant, every part of persisting object *O* that exists at that instant. The very notion of a temporal part has a temporal index built into it. By their very nature, temporal parts are objects that exist *at particular times*. So when we say that *P* is (a temporal) part of *O*, there is a disguised temporal index in that claim. What we are really saying is that there exists at t_1 some synchronic fusion *P*, and that fusion composes *O* at t_1. The difference between talking about synchronic fusions and temporal parts is that the notion of a synchronic fusion does not have a temporal location built into it.

So while the endurantist holds that *P* is part of *O* M-*simpliciter* just if there is some time *t* at which *P* is a part of *O*, so too the perdurantist holds that *P* is part of *O* M-*simpliciter* just if there is there is some time *t* at which *P* is part of O. Both endurantist and perdurantist agree that *P* is part of *O* M-*simpliciter*. For there *is* some time, namely t_1, at which *P* is part of *O* (though at t_1 *P* is, according to the perdurantist, a proper part of *O*, while for the endurantist it is an improper part of *O*). Similarly, while the perdurantist holds that *P* is part of *O* S-*simpliciter* at *t* just if *P* exists at *t* and is part of *O* M-*simpliciter*—*P* is the *t*-part of *O*—so too the endurantist holds that *P* is part of *O* S-*simpliciter* at *t* just if *P* exists at *t*, and *P* is part of *O* M-simpliciter. For the endurantist holds that *P* is part of *O* S-simpliciter at *t*, just if *P* constitutes *O* at *t*, and hence is an improper part of *O* at *t*.

Then the 'natural' reading of '*P* is part of *O*', which, according to the endurantist, is such that at t_1 the claim is true, and at t_2 is false, turns out to be the S-*simpliciter* reading. For at t_1 *P* is part of *O* S-*simpliciter*, and is not part of *O* S-*simpliciter* at t_2.

Ultimately then, both endurantist and perdurantist agree that there is one reading according to which the claim that *P* is part of *O* will be true at t_1 and false at t_2, and another reading according to which the claim will be true at all times. Both even agree as to what these readings are—the M-*simpliciter* and S-*simpliciter* readings. What they disagree on is which captures the natural, everyday sense in which we judge that some object has or lacks a part: the perdurantist holds that this is the M-*simpliciter* sense; the endurantist holds that it is the S-*simpliciter* sense. This, however, is mere semantic disagreement about the usage of words in the English language.

Both endurantist and perdurantist appeal to an under-defined or open-textured notion of a 'part' when one affirms, and the other denies, that'*P* is part of *O*' is true at t_2. For the three-dimensionalist notion of parthood has built into it the idea that no persisting object is at any time at which it exists, composed of non-present parts, while the perdurantist notion has built into it the idea that all persisting objects are at every time at which they exist, composed of parts some of which do not exist at those times.

These two senses of parthood are straightforwardly different senses—the former is the S-*simpliciter* sense, and the latter the M-*simpliciter* sense. Once we see this, however, we see that the apparent disagreement between the two views is just that—apparent. In the end both perdurantist and endurantist agree that it is tenselessly true that synchronic fusions exist at certain times, and at those times are related to persisting objects.

Let us then return to our definitions of endurance and perdurance. Recall from chapter three, that an object endures only if all of its parts S-*simpliciter* are present whenever it exists. I noted there that the END definition of endurance distinguishes endurance from perdurance because four-dimensionalists do not employ the M- and S-*simpliciter* apparatus, and because given that the S-*simpliciter* notion is supposed to capture the everyday sense of having a part, it is clear that the perdurantist will deny that persisting objects have all of their parts S-*simpliciter* present whenever they exist. This is nicely explained by the fact that as I have just argued above, the endurantist and perdurantist mean something different by 'part': when the perdurantist holds that persisting objects are composed of non-present parts she does not mean that they are composed of non-present parts S-*simpliciter,* but rather, that persisting objects have parts M-*simpliciter* which exist at times other than the present. Once we see that the perdurantist can happily accept the M- and S-*simpliciter* distinction, however, we can see that the perdurantist does not disagree that perduring objects have all of their parts S-*simpliciter* present whenever they exist. In essence, O has all of its parts present S-*simpliciter* at t_1, just if *O* has a temporal part (*O*-at-t_1), present at t_1. For the perdurantist, *O* has all of its parts S-*simpliciter* present at a time, just if the relevant synchronic fusion (the fusion that composes *O* at that time) exists at that time.

The same is true for the endurantist. An enduring object has all of its parts S-*simpliciter* present at t_1, just if *O* to has all of its parts t_1 ly at t_1: if *O* is constituted at t_1, by some synchronic fusion *F*. So a persisting object *O* has all of its parts present at time t_1 in the endurantist sense,[2] iff *O* has a temporal part present at t_1 in the perdurantist sense. Thus we have an assertibility mapping between the endurantist claim that '*O* is constituted by some synchronic fusion *F* at *t*,' and the perdurantist claim that '*O* has synchronic fusion *F* as a temporal part at *t*'. Similarly, the claim that '*O* is wholly present at *t*' maps onto the claim that '*O* has a temporal part present at *t*.' Then it follows that '*O* is wholly present at every time at which it exists' maps onto '*O* has a temporal part present at every time at which it exists.' And this latter claim is just the claim that *O* is a diachronic fusion: that it is a mereological fusion of temporal parts. Thus we can ultimately map '*O* is wholly present at every time at which it exists' onto '*O* is the mereological fusion of temporal parts.' And this is definition *P* of perdurance (chapter 3 p 73).

To be a mereological sum of temporal parts in the perdurantist sense, just is to be an enduring object that is constituted at every time at which it exists, by synchronic fusions that exist at those times, in the endurantist sense. Thus we see non-mereological universalism and mereological universalism as co-assertible and ultimately, I argue, inter-translatable. So there is no mereological axiom that is accepted by the perdurantist and rejected by the endurantist, at least, not once we clearly understand what is going on. It turns out (assuming we have a correct translation) that what it is to fuse synchronic fusions across time to create a diachronic fusion, is equivalent to those synchronic fusions constituting at times, some persisting object: diachronic fusions are enduring objects by another name.

With this assertibility mapping in mind we can now turn to the issue of whether we really have a correct translation between endurantism and perdurantism. As an interim step, we must first decide whether the assertibility mapping is truth preserving. To that end, in the following section I turn to consider each of the diagnostic criteria for a practical translation, beginning with explanatory equivalence.

1.2 Diagnosing a Practical Translation: Explanatory Equivalence

1.2.1 Temporary Intrinsics

It is touted as one of the primary advantages of perdurantism, that it allows one to hold that persisting objects instantiate properties *simpliciter*. The idea is that perdurantists analyse property instantiation in terms of there being some temporal part of a perduring object that instantiates a property *simpliciter*.[3] If *O* is a perduring object, then '*O* is red' is true at t_1 just if there is some temporal part of *O* that is red, that is, if *O*-at-t_1 is red *simpliciter*. (Since *O*-at-t_1 is instantaneous, if *O*-at-t_1 is red, then it is red *simpliciter*.) So perduring object *O* can be red at one time (t_1) and green at another (t_2) without any contradiction—for there is nothing contradictory about *O* having one part that is red, and another part that is green.

The 'traditional' endurantist, however, has considerably more trouble in explicating how enduring objects can instantiate properties *simpliciter*. In chapter three we considered this problem at length, which lead us eventually to embrace the idea that for the endurantist there are two different senses of having a property *simpliciter*: the M-*simpliciter* sense and the S-*simpliciter* sense. We will return to this distinction later. In brief, indexicalism analyses properties as disguised relations to times: enduring objects instantiate properties-at-times. Thus if *O* is red at t_1 and green at t_2, then *O* has the properties of being red-at-t_1 and green-at-t_2. The adverbialist, on the other hand, holds that it is the manner of instantiation of properties that is temporally relativised: thus properties are instantiated in particular temporal ways. *O* has the properties of being red $t_1$ly and green $t_2$ly. In general then, on the adverbial analysis, '*O* is red' is true at *t* just if *O* is red *t*ly. On the indexicalist analysis, '*O* is red' is true at *t* just if *O* is red-at-*t*. On the perdurantist analysis, '*O* is red' is true at *t* just if *O*-at-*t* is red.

It looks then, as though there is an assertibility mapping between '*O* is red *t*ly', '*O* is red-at-*t*' and '*O*-at-*t* is red'. But can this mapping be a practical translation given the difficulty the endurantist has making sense of the notion that properties are instantiated *simpliciter*? Well, the first thing to notice is that even for the perdurantist, perduring object *O* is red at t_1 in virtue of having some part at t_1 that is red *simpliciter*. *O* itself has no straightforward property of being red: the whole has the property of being red at t_1, just if it has the property of having-a-part-that-is-red-at-t_1. So *O* is not red *simpliciter*: only *O*-at-t_1 is red *simpliciter*. *O* is red in virtue of being related in a particular way to *O*-at-t_1, namely, by having *O*-at-t_1 as a temporal part. As a general schema, we might say that the perdurantist analyses property instantiation in terms of the existence of some synchronic fusion (*O*-at-t_1) that instantiates a property *simpliciter* (redness), and the existence of some persisting object (*O*), which is related in a particular way to that synchronic fusion.

The endurantist also holds that persisting objects are at all times at which they exist, related to synchronic fusions at those times. Thus she holds that there exists at t_1, some synchronic fusion which is at t_1 related to *O* in some manner. We can call this fusion *O*-at-t_1, though for the endurantist *O*-at-t_1 is not a temporal part of *O*, but rather, constitutes *O* at t_1. *O*-at-t_1 is red *simpliciter*. So the endurantist will hold that at t_1, enduring object *O* is constituted by an object—*O*-at-t_1—that is red *simpliciter*, and it is in virtue of this that *O* is red at t_1.[4] Schematically then, we might say that the endurantist analyses property instantiation in terms of the existence of some synchronic fusion (*O*-at-t_1) that instantiates a property *simpliciter* (redness), and the existence of some persisting object (O) which is related in a particular way to that synchronic fusion.

So for the endurantist, being red-at-t_1 or red $t_1$ly are reducible notions: O is red-at-t_1 or red $t_1$ly just if at t_1, O is constituted by some synchronic fusion that is red *simpliciter*. The endurantist appeals to talk of O being red-at-t_1 or red $t_1$ly when tenselessly analysing the properties that O has as a whole. Talk of being red-at-t_1 or red $t_1$ly is shorthand for talk of O being related in certain ways to something that is red *simpliciter* at t_1. And of course, the perdurantist could adopt exactly the same apparatus. She could hold that perduring object O tenselessly has the properties of being red-at-t_1 or red $t_1$ly just if at t_1, O is related to some synchronic fusion that is red *simpliciter*.

We can see then, that endurantist and perdurantist accounts of property instantiation are analogous. Both hold that only instantaneous objects (synchronic fusions) instantiate properties *simpliciter*, and that persisting objects instantiate properties at times by being related to those instantaneous objects in a particular manner at those times: by having them as temporal parts, or by being constituted by them at times.

So let us return briefly to the two notions of *simpliciter* that we explicated in chapter three. How does the idea of instantiating properties M- and S-*simpliciter* fit in with the endurantist's analysis of property instantiation? Clearly the endurantist has a huge advantage over the unitary three- dimensionalist, since for the former, instantiating properties in a temporally adverbialised manner ($t$ly) is not instantiating *irreducibly* temporally adverbialised properties. Still, it is true of O at every time at which it exists, that O is red $t_1$ly (including at t_2 when O is green). As we noted in chapter three, instantiating the property of being red $t_1$ly at a time, is not sufficient for an object to count as being red at that time. But what is the property of being red $t_1$ly? Well O is red $t_1$ly just if at t_1, O is constituted by some fusion—O-at-t_1—that is red *simpliciter*. So the tenseless property that O instantiates—being red $t_1$ly—is the property of being related at t_1, to a certain fusion (O-at-t_1) that is red *simpliciter*. Naturally, instantiating *this* property at a time is not sufficient for O to be red at that time. For when we say, at t_2, that O has the property of being red $t_1$ly, we are really saying that it is tenselessly true that 'at t_1, O has the property of being related at t_1, to a particular fusion that is red'.

So the sense in which O is red M-*simpliciter* in virtue of being red $t_1$ly, is not the sense in which O is red at all times at which it exists, but at some of those times fails to manifest redness. Rather, the sense in which O is red M-*simpliciter* is the sense in which it is tenselessly true of O, that at some time (t_1) O is related to a fusion (O-at-t_1) which is red *simpliciter*.

But O is only *red* at t_1, despite being red $t_1$ly at times other than t_1. For t_1 is the time at which O-at-t_1 exists. This is the sense in which O is red S-*simpliciter* at t_1: for at t_1 O has the property of being red $t_1$ly, and what it is to instantiate redness $t_1$ly at t_1, is for that time to be the one at which the relevant fusion—O-at-t_1—exists. This is why instantiating a property $t$ly at t is instantiating that property S-*simpliciter*: for instantiating a property P in the $_{now}$ly manner now, just is now being constituted by some synchronic fusion that is P *simpliciter*.

Exactly the same will be true if O is a perduring object. Given that the perdurantist uses the locution of O being *red* t_1*ly* to express the fact that it is tenselessly true of O, that it is red at t_1, then it follows that perduring object O will have the property of being red $t_1$ly at all times at which it exists. As with the endurantist, that property is the property of being related at t_1, to a certain fusion that is red *simpliciter* (having that fusion as a temporal part at t_1). Thus the distinction that the perdurantist draws between instantiating properties M- and S-*simpliciter*, is exactly analogous to the distinction that the endurantist draws. O is red M-*simpliciter* for the perdurantist just if there is some t at which O is related to a synchronic fusion that is red *simpliciter*. So O is red M-*simpliciter* in virtue of it being tenselessly true that at t_1, O is related to O-at-t_1 which is red *simpliciter*. O is red S-*simpliciter* at a time just if O is red $t$ly at t. For O is red S-*simpliciter* at a time just if that time is the one at which the relevant red fusion exists, that is, if the temporal part that exists at that time is red *simpliciter*.

So far then, we have shown that endurantism and perdurantism are equally explanatory with respect to the problem of temporary intrinsics: both are able to accommodate the idea that properties are instantiated *simpliciter*, and both explicate the idea of persisting objects instantiating temporary intrinsics, in terms of them being related at different times, to instantaneous objects that instantiate those properties *simpliciter*.

1.2.2 Temporary and Permanent Coincidence

This brings us to the puzzles of temporary and permanent coincidence. Consider first the temporary coincidence of the statue and the lump. Let us suppose that at t_1, the statue and the lump are materially coincident, while at t_2 they are not—suppose the statue exists at t_2 but the lump does not. Then the perdurantist will say that the statue and the lump are distinct perduring objects that overlap at t_1—that is, share a temporal part at t_1. Thus the synchronic fusion (call it F) that exists at t_1, is a temporal part of both the statue and the lump, explaining why at t_1, the statue and lump share the same intrinsic properties. Schematically then, the perdurantist analysis of temporary coincidence appeals to the fact that there exist multiple perduring objects both of which are related in the same manner at a time, to one and the same synchronic fusion.

The relation of sharing a temporal part at a time, has been described by Robinson as a four-dimensionalist version of constitution.[5] That is, Robinson argues that the perdurantist can make sense of the idea of constitution by holding that two perduring object O and O^* are (symmetrically) related by constitution at t just if at t O and O^* share a (maximal) temporal part. This is a perfectly good analysis of what a perdurantist might choose to mean by 'constitution', should she wish to incorporate this notion into her account. I see nothing wrong with calling this relation constitution, except that given our current terminology this would be unduly confusing. Instead, let us call this relation co-composition. Then:

> Co-composition: For any two or more perduring objects O and O^* that exist at a time t, O and O^* are co-composed at t just if (i) O has some maximal temporal part F at t, and (ii) O^* has some maximal temporal part F^* at t and (iii) F is identical to F^*.

Then the perdurantist will say that at t_1, the statue and the lump are co-composed, for the statue and the lump are each related in the same manner, to a single synchronic fusion (F) that exists at that time—they each have that fusion as a part. The endurantist will also analyse the relation between the statue and the lump in terms of the relation that both of them share to some third object, the synchronic fusion (F) that constitutes them both at t_1. In chapter four we defined this relation as the relation of co-constitution. Thus:

> Co-constitution: For any two or more enduring objects O and O^* that exist at a time t, O and O^* are co-constituted at t just if at (i) O is constituted by some fusion F at t, and (ii) O^* is constituted by some fusion F^* at t, and (iii) F is identical to F^*.

Hence the endurantist will hold that the statue and the clay are co-constituted at t_1, explaining why they share the same intrinsic properties—for each instantiates the intrinsic properties of the fusion that constitutes them both at t_1.

So both endurantist and perdurantist explicate temporary coincidence in an analogous manner, by appealing to the relation that each of the materially coincident persisting objects has at a time, with one and the same synchronic fusion. Both views therefore maintain that coinciding objects such as the statue and the lump are distinct in virtue of there being times (namely t_2)

at which they are not both related to one and the same synchronic fusion—times at which the objects are not co-composed or co-constituted.

Moreover, at t_1 the statue and the lump have distinct properties in virtue of the relevant synchronic fusion bearing different relational properties to each of the persisting objects. For the perdurantist, at t_1 fusion F has the property of being part of an object that exists at t_2 (the statue) and also has the property of being part of an object that does not exist at t_2 (the lump). Thus at t_1 the statue has the property of existing at t_2, and the lump lacks that property—we might say that the statue tenselessly has the property of existing $t_2$ly, while the lump tenselessly lacks that property. Similarly for the endurantist, synchronic fusion F has the property of constituting an object that exists at t_2, and constituting an object that does not exist at t_2. So the statue has the tenseless property of existing $t_2$ly, while the lump lacks that property. Hence, even at t_1, the statue and the lump have distinct properties.

What of Lumpl and Goliath? It is clear that both endurantist and perdurantist will say that Lumpl and Goliath are at all times at which they exist, related to the very same synchronic fusions: they are at all times co-constituted (according to the endurantist) or co-composed (according to the perdurantist). The real question is whether objects that are co-constituted or co-composed at every at time at which they exist, are distinct, identical, or contingently identical. Both endurantists and perdurantists can adopt any of these three options. Of course, for the perdurantist the option of holding that objects like Lumpl and Goliath permanently coincide yet are distinct, would require an amendment to the axiom according to which if O and O^* are fusions of all and only the same particulars, then O and O^* are identical. For Lumpl and Goliath share all and only the same temporal parts, and would by this axiom therefore be identical. Notice though, that an analogous axiom is a plausible one for the endurantist: if O and O^* are constituted by all and only the same synchronic fusions at times, then O and O^* are identical. If it is plausible that perduring O and O^*'s being composed of all of the same synchronic fusions means that they are identical, then it is also plausible that enduring O and O^*'s being constituted by all of the same synchronic fusions means that they are identical. Perdurantist and endurantist proponents of the view that Lumpl and Goliath are distinct, must, respectively, deny each of these plausible claims. In fact, as in the case of unitary three- and four-dimensionalism, I tend to think that the last option—contingent identity—is the best. For our purposes though, all that matters is that all of these explanatory strategies are open to the endurantist and the perdurantist.

Finally then, we can turn to consider a case of fission.

1.2.3 Fission

It is supposed to be a virtue of perdurantism that it, and only it, can reconcile all our various intuitions about fission. The perdurantist can reconcile both the claim that Riker survives transportation (the fission event) and that what matters in survival is identity. Recall that the perdurantist holds that there exist two perduring objects, one of which includes all of the temporal parts of one of the post-fission persons and of Riker, and one of which includes all of the temporal parts of the other post-fission person and of Riker—call the former R_1 and the latter R_2. Prior to fission, R_1 and R_2 share an extended temporal part. When we point to that part and call utter the name 'Riker' we are pointing to an object that is part of both R_1 and R_2. So to the extent that we think that 'Riker' names a persisting object of greater temporal extent than *just* that temporal part, it turns out that 'Riker' is ambiguous between referring to R_1, and referring to R_2. Given that both R_1 and R_2 exist prior to, and after fission, and since 'Riker' is ambiguous between referring to R_1 or R_2, the perdurantist can make sense of the claim that Riker survives fission. Moreover, since R_1 and R_2 are each self identical, the perdurantist can also say that what matters in survival is identity.

What should the endurantist say? In chapter one we considered a proposal due to Robinson, according to which the three-dimensionalist should embrace a view that is analogous to the one adopted by the perdurantist: she should hold that prior to fission there exist two materially coincident persons R_1 and R_2 to which the name 'Riker' ambiguously refers.[6] Prior to fission R_1 and R_2 are related by constitution, but post-fission R_1 and R_2 'split apart' and cease to be related by constitution. I argued, in chapter one, that this approach has some difficulties in the context of traditional three-dimensionalism. Absent a commitment to some sort of backwards causation it implies that there exist multiple coincident objects (in this case persons) regardless of whether any fission event occurs or not: so long as fission is possible then multiple coincident objects must exist, and where fission does not occur, those objects will permanently coincide.

It is easy to see how Robinson's account of fission can be modified to suit the endurantist. The endurantist will *not* hold that prior to fission, R_1 and R_2 are related by constitution. Rather, she will hold that prior to fission there exist certain synchronic fusions, which, at each time at which they exist, constitute multiple enduring objects. One of those objects is R_2, and the other is R_1. Prior to fission, R_1 and R_2 are constituted by the very same fusions: they are co-constituted. After fission, R_1 and R_2 are constituted by distinct fusions. Thus both R_1 and R_2 survive fission, and since 'Riker' refers ambiguously to R_1 and R_2, this is the sense in which Riker too survives fission. The question is, however, whether this account is equally at the mercy of the objection that I outlined in chapter one. It seems so.

In chapter five I considered unitary three- and four-dimensionalist analyses of fission, but I neglected to consider whether one or both of those analyses might also fall prey to the objection in question. I turn to that issue now. It appears that unitary three-dimensionalism will be faced with the same difficulty as endurantism: if future fission events do not determine how many objects exist in the present, then there must in every case in which fission is possible, exist multiple coincident objects. It is not so clear that unitary *four-dimensionalism* faces the same problem. On the one hand, unitary four-dimensionalists do not hold, as the perdurantist does, that there exists a single object prior to fission (an extended temporal part) and the issue is simply whether that object is part of one or two persisting objects. Rather, the unitary four-dimensionalist holds that there exist persisting fusions, which compile at times, persisting non-fusions (such as R_1 and R_2). So we might be tempted to think that how many coincident objects exist at a time, that is, how many persisting non-fusions a fusion compiles at a time, is not something that is determined by some future event. If we think that, then we think that the unitary four-dimensionalist has the same problem as the unitary three-dimensionalist.

Plausibly though, the four-dimensionalist might argue that future facts are, without any sort of backwards causation, relevant in determining how many objects coincide at a time. Since terduring objects are four-dimensional, determining how many exist at a time will involve looking at both past and future times. In order to know how many distinct non-fusions some fusion compiles at a time, we need to trace each of those objects both backwards and forwards in time—only by examining the *entire* terduring objects O and O^* can we determine whether they are distinct. For we need to know whether, given that O and O^* coincide in the present, they were distinct in the past (they have fused) and whether they will be distinct in the future (whether fission will occur). Thus the unitary four-dimensionalist can argue that because terduring objects are four-dimensional, the number of coincident terduring objects at a time will depend on both past and future facts, though not in virtue of any backwards causation.

If we assume that both unitary and non-unitary four-dimensionalism are immune to this objection, then have we discovered an explanatory advantage for the four-dimensionalist? Consider the following. A case in which fission could have occurred but does not, is essentially analogous to a case of permanent coincidence. Suppose the three-dimensionalist (unitary or non-unitary) thinks that Lumpl and Goliath are distinct and necessarily so. Then she thinks that two distinct objects permanently coincide in the actual world, and thinks that they are distinct in virtue of the fact that there are counterfactual worlds, including w_1, say, in which those objects do not coincide permanently. If Lumpl and Goliath were the same *kinds* of object, then this temporary coincidence, in $w_{1,}$ would be an instance of fission. But if Lumpl and Goliath are actually distinct in virtue of having distinct counterparts in w_1, then by the same reasoning we ought to think that if Riker had not actually undergone fission, then in our world R_1 and R_2 are distinct *permanently* coinciding objects in virtue of having distinct counterparts w_1. If we rule out *contingent* identity, as this view does, then it cannot be that R_1 and R_2 (in w_1) share a single counterpart in this world (Riker). Rather each must have a distinct person-counterpart in this world—R_1 and R_2—it is just that these counterparts coincide permanently.

So if the three-dimensionalist thinks that in our world Lumpl and Goliath are distinct permanently coinciding objects, then by parity of reasoning she ought also to think that has fission not actually occurred, then R_1 and R_2 would have been distinct permanently coinciding objects. That simply follows from the claim that if objects are distinct, then they are necessarily distinct. And since there are counterfactual worlds in which many hundreds of post-fission 'Rikers' are created, it follows that all of those distinct persons are also distinct in our world (albeit permanently coinciding). Hence any name 'N' is massively ambiguous between referring to many permanently coinciding objects.

Notice though, that it is not just the three-dimensionalist who will say this: the four-dimensionalist who holds that Lumpl and Goliath are distinct permanently coinciding objects must agree. So if this ontological profligacy is problematic for the three-dimensionalist, it is equally so for the four-dimensionalist. Perhaps this is reason to reject the view that Lumpl and Goliath are distinct. Of course, both three- and four-dimensionalists can point out that where objects of the same kind permanently coincide, it can be perfectly proper to describe this situation in natural language, as being one in which only one object exists. For they can hold that in natural language, we frequently do not count by identity. Moreover, both can point out that even in the case of actual objects that have no counterparts that undergo fission, it will still be the case, given unrestricted composition, that such objects temporarily coincide with objects of the same kind. Even if Riker has no fission counterparts, it is still true (given unrestricted composition) that for every sub-interval over which Riker exists, there exists some persisting object that coincides with Riker during that interval, and which exists only during that interval. Perhaps these considerations ameliorate somewhat the counterintuitive claim that there exist many permanently coinciding objects, some of which are of the same kind.

What though, is the three-dimensionalist (unitary or non-unitary) to say if she holds that permanently coinciding objects such as Lumpl and Goliath are contingently identical? By analogous reasoning to that above, she should conclude that if fission does not occur in the actual world, then R_1 and R_2 are contingently identical. But to say this is to say no more than that there exists a single individual in the actual world, which has multiple counterparts in counterfactual worlds. It is to say that there exists a single person—Riker—in the actual world, and that in some counterfactual world (w_1), Riker has two counterparts, R_1 and R_2. So in a world in which fission does not occur, there exists only a single individual, not multiple individuals that coincide permanently.

This, however, seems to run foul of the claim that for the three-dimensionalist, since objects are wholly present whenever they exist, there must be some fact of the matter as to how many objects exist and coincide at a particular time, and that fact of the matter should supervene on the intrinsic properties of the objects at that time. That is, what determines how many objects coincide at a time should not be some future facts that somehow retrospectively make it the case that multiple objects exist, or fail to exit at an earlier time. Given a contingent identity account of permanent coincidence, however, this is precisely what seems to be required.

We know that four-dimensionalists can appeal to future facts in a way that does not appeal to some sort of backwards causation. But why should future facts determine how many *three*-dimensional objects exist at an earlier time? I have noted throughout that even at the times at which temporarily coinciding objects such as Statue and Clay coincide, they have different intrinsic properties. Suppose that Statue and Clay coincide at t_1, and at t_2 do not coincide. Suppose that at t_1 Clay and Statue are blue, and at t_2, Clay is red and Statue is green. At t_1 Clay and Statue share all of the intrinsic properties of the fusion that constitutes them at that time. That is, at t_1 they share all of the same $t_1$ly properties. For the $t_1$ly properties of each, supervene on the properties of the fusion that constitutes each of them at t_1. But since both Statue and Clay exist at times other than t_1, at t_1 they have more properties than just their $t_1$ly properties. At t_1, Statue has the property of being green $t_2$ly, while Clay has the property of being red $t_2$ly. And just as being blue $t_1$ly is supposed to be an intrinsic property of both Statue and Clay, so being green $t_2$ly is supposed to be an intrinsic property of Statue, and being red $t_2$ly an intrinsic property of Clay. But the intrinsic property of being green $t_2$ly which Statue instantiates at t_1, is a property whose truth maker exists at t_2: Statue is green $t_2$ly at t_1, in virtue of being constituted, at t_2, by a fusion that is green. So even though it is an intrinsic property of

Statue that it is green $t_2$ly at t_1, the truth make for that claim is nevertheless extrinsic to t_1.

Presumably the three-dimensionalist will say something similar about cases of fission. If there is some fission event, then prior to fission it will be true of R_1 and R_2 that they each have different properties. Prior to fission, R_1 and R_2 have different properties $t_n$ly, where t_n refers to any time after the fission event, at which R_1 and/or R_2 exists. And it is in virtue of having these different properties prior to fission, that R_1 and R_2 are distinct prior to fission. Of course, the different properties that each have prior to fission, are properties whose truth makers are extrinsic to that time. But they are nevertheless properties that each object has *at those times*. So just as it is not the case that the green fusion that exists at t_2 and constitutes Statue, retrospectively *causes* it to be the case that Statue is green $t_2$ly at t_1, so too it is not the case that a future fission event retrospectively causes it to be the case that multiple coincident persons exist prior to that event. Rather, it is that distinct objects are distinct in virtue of having different properties prior to fission, although the truth makers for those properties exist in the future. Then if no fission event occurs, there are not two permanently coinciding objects,. Rather, if fission does not occur there exists just one individual with one set of properties, but this individual has multiple counterparts in some worlds, worlds in which fission occurs.

Thus both the three- and four-dimensionalist who embrace contingent identity will appeal in some manner to future facts. Since unitary four-dimensionalists hold, with unitary three-dimensionalists, that properties are temporally adverbialised, they too hold that it is tenselessly true of a terduring statue that it is green $t_2$ly at t_1. Ultimately, for the unitary four-dimensionalist, what distinguishes the terduring Statue and Clay at t_1, is that although each is compiled by the same fusion at t_1, each has distinct properties: each has distinct $t_2$ly properties. The same is true in cases of fission: R_1 and R_2 have different tenseless properties ($t$ly properties) and the truth makers for those different properties exist at times after fission. So for both unitary three- and four-dimensionalists, the mechanism that allows each to say that prior to fission there exist distinct objects that temporarily coincide, is an appeal to the tenseless properties of the objects in question, properties that the objects have at every time at which they exist, but whose truth makers may exist at other times.

Moreover, as we have seen, non-unitary four-dimensionalists (perdurantists) also hold that the sense in which perduring objects tenselessly have properties—have them M-*simpliciter*—is the sense in which it is tenselessly true that they have some part that is red *simpliciter*. So the truth maker for this claim will be extrinsic to many of the times at which the claim is true. Thus just as for the non-unitary three-dimensionalist, claims about the distinctness of R_1 and R_2 prior to fission in some sense supervene on future facts about R_1 and R_2 post-fission, the same is true for the non-unitary four-dimensionalist—for prior to fission it is tenselessly true, for the four-dimensionalist, that R_1 and R_2 are distinct, and have different properties M-*simpliciter*. So we can see that those three- and four-dimensionalists who embrace contingent identity explicate fission in an analogous manner. Ultimately, whether they construe cases of putative permanent coincidence in terms of actual distinctness or contingent identity, analogous versions of three- and four-dimensionalism bring the same explanatory resources to the puzzle of fission.

At this point then, I set the issue of fission to rest. In the following chapter, however, I go on to discuss some issues pertaining to time travel. The nature of that discussion sheds further light on the puzzle of fission and thus I will briefly return to some of these issues in chapter six.

So far we have shown that endurantism and perdurantism deal with the puzzles of persistence in the same manner. Thus we have good reason to suppose that they are equally explanatory: they meet the first criterion of a practical translation. Indeed, it is not merely that each is able to explain the same phenomena. For that would be true even if they explained the same phenomena, but in a wildly different manner. What is noteworthy is that the explanations do not look different. So there really is a robust sense in which they are explanatorily equivalent.

In the next sub-section I move on to consider another diagnostic criterion of a practical translation—the principle of charity, before in section two taking a short detour to consider some objections to perdurantism. Finally, in section three I consider whether the practical translations that we have developed are also correct translations: have we provided good reason to think that three- and four-dimensionalism are metaphysically equivalent?

1.3 The Principle of Charity

In the previous chapter I argued that the principles of charity and humanity provide still further reason to suppose that the assertibility mapping between unitary three- and four-dimensionalism is truth preserving. For, I argued, there does not seem to be any information—empirical, logical or metaphysical—that, say, the (unitary) three-dimensionalist could come to know and in virtue of which she would come to realise that by her own lights, three-dimensionalism is false. So too the same is true *mutatis mutandis* for the (unitary) four-dimensionalist.

Equally, the same can surely be said for the non-unitary three- and four-dimensionalist—the endurantist and the perdurantist. Here too, one can imagine no piece of information that would 'remove the scales from the eyes' of the proponent of either theory such that she would recognise her view as mistaken. Given this, the principle of humanity tells us that we ought to treat the utterances of the proponents of both theories as expressing truths. So we ought to think that our assertibility mapping is truth preserving: we ought to think that we have a practical translation.

So far then, we have a coherent assertibility mapping between theories that are equally explanatory, and where considerations of interpretation militate in favour of seeing those theories as both expressing truth. Hence we have good reason to suppose that we have a practical translation. Before I move on to argue that these practical translations are correct translations, however, it is first necessary to make a small detour through the landscape of modal properties and contingent identity. For it has been argued that four-dimensionalists are ultimately forced to reject trans-world identity in favour of counterpart theory. If this is so, and if three-dimensionalists are not also forced to take this metaphysical course, this would be evidence that three- and four-dimensionalism are not equivalent: one has resources that the other does not. I turn to examine these issues in the following section.

2. MODAL PROPERTIES AND CONTINGENT IDENTITY

I have already noted that both three- and four-dimensionalism in their unitary and non-unitary guises have the same three options open to them with respect to cases of permanent coincidence such as that illustrated by the story of Lumpl and Goliath. They have the options of holding that Lumpl and Goliath are identical and necessarily so, that they are distinct and necessarily so, or that they are contingently identical. There are, however, a number of arguments that purport to show that given some plausible assumptions about modal properties, perdurantism is faced with some difficulties that can only be successfully resolved by recourse to counterpart theory and contingent identity. If that is so, then my earlier claims about Lumpl and Goliath are false: there is only one possible response that the perdurantist can make, and this is to embrace contingent identity.

In what follows I examine the arguments that purport to show that perdurantists should accept some version of contingent identity, and I consider whether these arguments apply equally to non-unitary versions of three-dimensionalism. I argue that indeed they do. If the arguments succeed, then they show that both the perdurantist and the endurantist ought to embrace contingent identity. While this may not be good news for trans-world identity theorists, it is no threat to the claim that three- and four-dimensionalism are equivalent. I will argue, however, that in fact neither endurantist nor perdurantist need accept counterpart theory as a response to these type of arguments. For neither endurantist nor perdurantist need accept that any persisting objects are modally inductile, and rejecting this idea allows both to avoid the counterintuitive consequences of the arguments in question.

2.1 Modal Inductility Arguments

There are a number of related arguments that aim to show that perdurantism cannot accommodate our firm intuitions about some modal properties of persisting objects—or at least, that absent an appeal to counterpart theory, perdurantists cannot accommodate these intuitions. Both Van Inwagen[7] and Howard-Snyder[8] construct essentially the same argument. Suppose Descartes lived for exactly 54 years. Consider the modal predicate 'possibly dies before age 54' and call this property P^*. Descartes either has P^* or he lacks it. Plausibly, he has this property, since otherwise it is necessary that he lived for at least as long as he did. Given perdurantism, among Descartes' temporal parts is an improper temporal part, namely that part that includes all and only Descartes' proper temporal parts. Call that temporal part D^*. By P^* we know that Descartes could have had a shorter life span than he did. If we suppose that D^* is identical to Descartes (since they share all and only the same temporal parts) then it follows that D^* also has the P^* property of possibly being shorter than it is. Then consider another of Descartes' temporal parts, the part that began at his birth and ended ten years prior to his death. Call this D^-. If Descartes had died ten years earlier than he did, then D^- would have existed unchanged.

So consider a world w in which Descartes dies ten years earlier than he does actually. Should we think that D^* exists or fails to exist in w? On the one hand it seems we should think that D^* fails to exist. The essential property of a temporal part is its temporal extent, and in w, D^* would have a different temporal extent. Moreover, if we thought that D^* did exist in w, then we would have to conclude that D^* is identical to D^-, since they both share all the same temporal parts. Then D^* would be identical with D^-, despite the fact that D^- has the property of possibly being distinct from D^* (since they are distinct in the actual world). Then D^* and D^- do not share all the same properties, since D^* does not have the property of possibly being distinct from D^*. On the other hand there seems to be good reason to think that D^* does exist in w. For Descartes exists in w, and D^* is Descartes. Either way, a commitment to perdurantism seems to lead to an incoherent consequence.

This argument rests on the following premises: (1) at least some persisting objects are modally ductile—they could have had a different temporal extent from the one they actually have (2) temporal parts have their extent essentially (3) temporal parts are modally inductile (4) improper temporal parts are necessarily identical to the persisting objects of which they are improper parts (5) for any persisting object O and sub-interval I over which O exists, there exists some persisting object O^* that exists during and only during I and which materially coincides with O whenever it exists.

Given these five premises, it is easy to see why van Inwagen[9] recommends a rejection of (4) in favour of some sort of counterpart theory. One could reject (5) and hence maintain, for instance, that only instantaneous temporal parts and improper temporal parts exist. Thus D^- does not exist. But this seems arbitrary and implausible. Similarly, a rejection of (1) seems implausible, since we are surely committed to Descartes being modally ductile. So the only possibility, other than counterpart theory, is to argue that temporal parts do not have their temporal extent essentially, and hence are not modally inductile. We will consider this option shortly. First, however, let us see how an adoption of counterpart theory would resolve the apparent contradiction.

2.1.1 Counterpart Theory and Modal Ductility

Counterpart theory is the view that a single object in the actual world, may have multiple counterparts in counterfactual worlds depending on which counterpart relation we track in each of those worlds. It is this apparatus that allows us to make sense of contingent identity, whereby, say, Lumpl and Goliath are identical in the actual world—the designation 'Lumpl' picks out the same actual object as the designation 'Goliath'—but the lump counterpart relation picks out a different object in some counterfactual world w than does the statue counterpart relation. Thus Lumpl has lump counterparts that are distinct from the statue counterparts of Goliath.

Appeal to counterpart theory and thus a rejection of premise (4) allows us to resolve the puzzle of Descartes by noting that Descartes and D^* invoke different counterpart relations: associated with Descartes is a personal counterpart relation that picks out other relevantly similar persons in other worlds. Associated with D^* is a temporal part counterpart relation that picks out relevantly similar temporal parts in other worlds. Since D^* has its temporal extent essentially, an object in a possible world counts as a counterpart of D^* only if it has the same temporal extent as D^* does actually. An object in a possible world counts as a Descartes' counterpart only if the personal counterpart relation picks out a person in that world that is a maximal aggregate of temporal parts of a person united in an appropriate manner—whichever manner preserves Descartes' identity across time.

'Descartes' and 'D^*' pick out the same four-dimensional object in the actual world. But when that object is picked out by the designation 'D^*', it invokes a temporal part counterpart relation, such that it is the case that D^* could not have had a different temporal extent. When that same object is picked out by the designation 'Descartes' then a different counterpart relation is invoked—the personal identity relation—such that Descartes could have had a different temporal extent.

What then, are we to say about world w? It seems that in w we are faced with a dilemma. If D^* exists in w then it seems we should say that D^* both is and is not identical to D^-. What reason do we have to think that D^* exists in w? According to Howard-Snyder, because Descartes exists in w, and Descartes is identical to D^*. It is easy to see, however, why given counterpart theory this is a mistake. It is true that Descartes exists in w (more precisely, it is true that a Descartes' counterpart exists in w). But it is not true that D^* exists in w, for D^* has its temporal extent essentially and hence there cannot exist a D^* counterpart in w. In the actual world, Descartes is contingently identical to D^*. In w, Descartes is contingently identical to D^-. But since D^* and D^- both invoke temporal part counterpart relations which preserve temporal extent, it will never be true that D^* is contingently identical to D^-, and since counterpart theory does not preserve transitivity of contingent identity, there is no pressure to identify D^* with D^- just because each is contingently identical with Descartes in different worlds.

Thus our intuitions are preserved. We get to say both that temporal parts have their extent essentially and that many persisting objects, such as persons, do not have their extent essentially, and yet we also get to say that persisting objects and their improper temporal parts are contingently identical.

2.2 Endurantists and Modal Inductility

So far then, we have seen that the perdurantist can answer this objection—what we might call *the modal inductility argument*, by appeal to counterpart theory. What of endurantists? Are endurantists immune to this argument?

I see no reason to think that they are. After all, endurantists, like perdurantists, hold that for any persisting object O and sub-interval I during which O exists, there exists some enduring object O^* that exists during and only during that interval, and which during that interval materially coincides with O. The endurantist too holds that many enduring objects are modally ductile: persons, trees, dogs and so forth. But what of the various materially coincident enduring objects with which trees and dogs are co-constituted at times?

At least *prima facie*, it looks as if whatever reason we might have for thinking that temporal parts have their extent essentially, will equally apply to these enduring objects. (After all, we can define each in terms of them being the objects that overlap persisting objects at certain times, and which exist during and only during those times). And if that is the case, then the argument from modal inductility has as much force against the endurantist as it does against the perdurantist. For notice that the motivation for the perdurantist to think that D^* is identical to Descartes, is that D^* is an improper part of Descartes, and various mereological axioms tell us that objects and their improper parts are identical. The endurantist, though, will also hold that D^* is an improper part of Descartes. Most endurantists say that for any two enduring objects O and O^*, O is an improper part of O^* at a time t, just if at tO and O^* materially coincide. So it is not a large leap to move from the claim that D^* is an improper part of Descartes at every time at which D^* and Descartes exist, to the tenseless claim that D^* is an improper part of Descartes *simpliciter*. Then by parity of reasoning with the perdurantist, it follows that if improper parts of objects are identical to those objects, then enduring objects Descartes and D^* are identical and necessarily so. If that is so, then the argument from modal inductility takes force, and endurantist and perdurantist alike find themselves in the same modal inductility boat.

Of course, rejecting premise (4) need not amount to adopting counterpart theory and hence contingent identity. 'Traditional' three-dimensionalists are more apt to think that D^* and Descartes (or more usually Lumpl and Goliath) are distinct and necessarily so, rather than that they are contingently identical. That approach too resolves the modal inductility problem. For then Descartes, D^* and D^- are all distinct objects, some of which permanently coincide in some worlds in which they exist. Notice that as I suggested previously in this chapter, the four-dimensionalist can also embrace this view. Just as the three-dimensionalist embraces this view at the cost of holding that objects and their proper parts are distinct, and thus that frequently objects of the same kind permanently coincide, (such as R_1 and R_2) so the four-dimensionalist who embraces that view incurs the same costs. For those who reject premise (4), these costs may militate in favour of adopting the contingent identity thesis, but this latter thesis is not the only option. The real point is that the modal inductility problem is equally a problem for three- and four-dimensionalists who wish to embrace premise (4).

2.2.1 The Modal Ductility of Persisting Objects

In fact, I think the moral of this story is that we should be wary of the claim that any persisting object has its temporal extent essentially. Why should we think that temporal parts have their extent essentially? One reason might be the way in which a maximal temporal part is frequently defined. Recall that in chapter three (p 72) we defined extended temporal parts as those objects that overlap persisting objects at, and only at, certain times. Given this definition of an extended temporal part, it is perhaps not surprising that we might conclude that the essential feature of such an object is its temporal extent. And, of course, we could define enduring objects in the same way, by noting that they are those persisting objects that exist during and only during certain times, and which at each of those times are constituted by certain fusions. Then when we are considering some enduring object that is co-constituted with another enduring object at certain times, we are considering an object that exists during and only during certain times, and which at those times overlaps every spatial part of the object with which it is co- constituted. That is, we could say that:

> An enduring co-constituter of x during T is an object that exists at all and only times in T, is part of x at every time during T, and at every moment in T overlaps everything that is part of x at that moment.

(Notice that this second clause is still true of enduring objects, since enduring co-constituters are parts at times of the objects that they co-constitute, it is just that they are improper parts at those times.)

However, from chapter four onwards this is not how I have presented endurantism or perdurantism. Instead, I have emphasised the fact that both endurantists and perdurantists are committed to the existence of various synchronic fusions, and that in addition, perdurantists hold that there exist various diachronic fusions of these synchronic fusions, while endurantists hold that there exist various enduring objects that are constituted by these fusions at times. This represents a subtly different way of thinking of things. Diachronic fusions are fusions of synchronic fusions. Thus suppose that there exists some diachronic fusion D that is the fusion of synchronic fusions S_1, S_2, S_3 and S_4 which exists at times t_1, t_2, t_3 and t_4. Then D will have as a proper part, some temporally smaller diachronic fusion D^-, which is the fusion of S_1, S_2 and S_3. Hence D^- is a temporally extended object that overlaps D between t_1 and t_3, (call this interval T) and exists during and only during T. But it is not *definitive* of D^- that it has T as its temporal extent. D^- is not defined as the object that exists during and only during T and overlaps D at those times. D^- is simply the diachronic fusion of S_1, S_2 and S_3, just as D is the fusion of $S_1 \ldots S_4$. The fact that D^- is a proper temporal part of D in no way implies that we ought think that D's temporal extent is essential, while D's extent is not.

Indeed, once we start to think about this a little, the above can hardly be surprising. For suppose that we thought that only maximally R-related objects are modally ductile—where some object is maximally R-related just if is the maximal object that preserves R-ness. Then we might have thought that Descartes is the four-dimensional object that is maximally person-related. Hence Descartes does not have his temporal extent essentially, and he is not modally inductile. But, we might have suggested, all of Descartes' proper temporal parts are, by definition, not maximally person-related: so these parts, do have their temporal extent essentially, for what is definitive of these parts is that they overlap, at times, some maximally person-related object.

But of course, this just won't work. It would allow us to say that Descartes and D^* are identical, since D^* would not have its temporal extent essentially—D^* is, after all, also a maximally person-related object. So too, we could say that Descartes is identical to D^- in w, since in w D^- does not have its extent essentially. This is where it all falls down. For D^- in the actual world has its temporal extent essentially, since in our world D^- is a proper part of Descartes and D^*. But in w, D^- is not a proper part of Descartes, and does not have its extent essentially. But D^-cannot both have, and fail to have its extent essentially.

The problem is that it makes no sense to think that wholes (ie. objects that are maximally R-related), are ductile, while also thinking that extended temporal parts that are proper parts of those wholes are inductile. For any maximally R-related object, such as the maximally person-related Descartes, will be some proper part of some other maximally R-related object (where 'R' refers to a different relation in each case). Descartes, for instance, is a proper part of some maximally body-related object that includes Descartes' corpse. Indeed, there are any number of maximally R-related things with which Descartes is continuous. But none of this provides us with reason to suppose that Descartes has his temporal extent essentially: being an extended temporal part of some larger persisting object is not a reason to conclude that that extended temporal part has its extent essentially.

Moreover, there seem clear cases where we want to say that extended temporal parts do not have their extents essentially. We may want to say, for instance, that Descartes' adolescence could have been longer, or shorter. Yet Descartes' adolescence is an extended temporal part of Descartes.

As I see it, there is no principled reason to suppose that some temporally extended objects, like Descartes, fail to have their temporal extent essentially, while other temporally extended objects, like D^*, or D^-, or Descartes' adolescence, do have their temporal extent essentially. Nothing in perdurantism entails that this is so. The perdurantist can, of course, maintain that synchronic fusions have their temporal extent, (or lack thereof), essentially. But it does not follow from the fact that synchronic fusions are essentially instantaneous, that the temporally extended objects that have those synchronic fusions as parts have their extent essentially, any more than it would follow from the fact that the most fundamental particulars have their spatial extent essentially, that the mereological composites that are composed of those simples have their spatial extent essentially. Hence the perdurantist can maintain that while no synchronic fusion is ever trans-world identical to some temporally extended object, nevertheless, one and the same trans-world temporally extended object has different temporal extents in different worlds in virtue of being composed of different synchronic fusions in different worlds.

Naturally, the endurantist can adopt an analogous view: she can hold that while the fundamental synchronic fusions are essentially instantaneous, the various enduring objects that are constituted by those fusions at times, do not have their temporal extent essentially. Rather, in different worlds one and the same enduring object may be constituted by different synchronic fusions and hence may endure for a longer or shorter period of time.

Thus for those endurantists and perdurantists who think that some persisting objects have their temporal extent essentially, resolution of the modal inductility argument requires a rejection of premise (4), and hence requires either holding that Descartes and D^* are contingently identical, or that they are distinct and necessarily so. For those endurantists and perdurantists who hold that no persisting objects have an essential temporal extent, the modal inductility argument can be resolved without recourse to contingent identity. While my preference lies with the latter view, what is important is that either way, both endurantist and perdurantist have equal resources when dealing with modal properties, contingent identity and modal (in)ductility.

3. A CORRECT TRANSLATION?

So far, hopefully I have shown both that there is an assertibility mapping between analogous versions of unitary three and four-dimensionalism, and between analogous versions of non-unitary three and four-dimensionalism, and further, that that mapping counts as a practical translation: it preserves truth. For I have argued that each of the diagnostic criteria of a practical translation are met: these theories are empirically equivalent, equally explanatory, and the principle of charity/humanity tells us that we should interpret each as being true. Given that these diagnostic criteria are met, we have good reason to suppose that there exists a practical translation.

But given that the diagnostic criteria provide necessary but insufficient conditions for some assertibility mapping to count as a correct translation, that we have a practical translation does not entail that we have a correct translation. For it is not sufficient for a mapping to count as a correct translation that the mapping preserves truth: it must preserve truth in virtue of the theories in question having the same truth makers.

In chapter one I raised the issue of whether we are licensed to conclude, from the existence of a practical translation, that there also exists a correct translation. I argued that there is such a reason. If we have a practical translation between two theories, then the only way that those theories could fail to be correctly inter-translatable and hence equivalent, is if there were some extra metaphysical facts in virtue of which one theory is true and the other false. Such metaphysical facts would, however, need to be unobservable, explanatorily redundant truth makers: for we already know that the theories in question are equally explanatory, empirically equivalent, and that the translation preserves truth. So is there any reason to suppose, with respect to any of the analogous theories of three- and four-dimensionalism, that there exist any such extra unobservable facts?

I think not. Of course, those dogmatists who insist that no versions of three- or four-dimensionalism are equivalent, will staunchly persist in believing that there do exist such facts. Absent this assertion, however—that is, absent the desire to show that these theories are not equivalent—I see no reason to posit the existence of such facts. For unlike a theory of modal realism, or a theory about black holes, each of which posit the existence of certain unobservable facts as an explanatorily salient part of the body of the theory, no such claim seems compelling with respect to three- or four-dimensionalism. Or at least, no such claim seems compelling once we compare analogous versions of three- and four-dimensionalism and see that they meet each of the diagnostic criteria. *Prima facie* we might have thought that the perdurantist's positing of temporal parts is an example of positing the existence of some unobservable yet explanatorily salient element. What we have seen in these last two chapters is that this is a mistake. Perdurantists do posit the existence of temporal parts in order to explain certain phenomena, but those same phenomena can be explained by the endurantist who also appeals to the existence of the very same objects—synchronic fusions—but holds that those synchronic fusions constitute, instead of being parts of, persisting objects at times.

For both unitary and non-unitary three- and four-dimensionalists, the debate ultimately hangs on whether or not persisting objects are strictly identical across time. If three- and four-dimensionalism are not equivalent, then it must be in virtue of the existence of some unobservable metaphysical fact regarding whether or not persisting objects are strictly identical across time. It must come down to a brute claim that such a fact obtains, or fails to obtain: for nothing else can distinguish these theories. Yet neither theory posits any such fact. Of course, the three-dimensionalist holds that persisting objects *are* strictly identical across time, while the four-dimensionalist denies this. But neither implies that whether or not this fact obtains ought to be both unobservable and explanatorily redundant. Rather, both theories think that this is the crucial metaphysical and explanatory *difference* between their theories. Neither theory posits the existence of such a fact qua unobservable explanatorily redundant fact. This latter is required only once we see that we have a practical translation between three- and four-dimensionalism, and then it begins to look a lot like an *ad hoc* measure whose only purpose is dogmatically to maintain that the theories are not metaphysically equivalent.

Yet as soon as either theorist ceases to appeal to a brute metaphysical fact, it becomes plausible indeed to think that theories that are practically inter-translatable are also correctly inter-translatable. As we noted in chapter two, the best explanation for the existence of some assertibility mapping that preserves truth, is that the mapping is a correct one. So it is the fact that three- and four-dimensionalism are correctly inter-translatable and hence metaphysically equivalent, that explains why they are co-assertible, empirically equivalent, explanatorily equivalent, truth preserving and so forth. For if these theories did not have the same truth makers, then it would be a complete mystery why all of those features obtain. It is only by noting that the theories are equivalent, that we can explain why this is the case.

4. A PREFERABLE THEORY OF PERSISTENCE?

My hope is that the discussion in this and the previous chapter shows that there is good reason to think that analogous versions of three- and four-dimensionalism are metaphysically equivalent. In arguing for this conclusion, I have developed a number of novel theories of persistence that have been examined in some detail. Consideration of these theories has not only provided an account of their various explanatory features, but in addition, has revealed what sorts of metaphysical commitments each theory makes. A question that arises from a consideration of these views is—which is the most plausible? I want to briefly consider this question shortly. But first we might wonder whether, given the transitivity of equivalence, there is really any question to be asked.

4.1 The Transitivity of Equivalence

In chapter five I argued that the unitary three-dimensionalist's notion of constitution maps onto the unitary four-dimensionalist's notion of compilation. Thus we can translate talk about constitution, into talk about compilation and *vice versa*. In this chapter I argued that the non-unitary three-dimensionalist's (endurantist's) notion of constitution, maps onto the non-unitary four-dimensionalist's (perdurantist's) notion of having a temporal part at a time. Hence we can translate talk of having some temporal part at a time, into talk of being constituted at that time, by some synchronic fusion. So each version of three- and four-dimensionalism is inter-translatable.

But if what I have said is true, then something more also seems to be true. For the notion of constitution employed by the unitary three-dimensionalist, is the same as that employed by the non-unitary three-dimensionalist. Then given the transitivity of identity, it turns out that non-unitary three- and four-dimensionalism are equivalent to unitary three- and four-dimensionalism (more precisely, that restricted versions of each are equivalent, as are unrestricted versions of each). Thus perdurantism is equivalent to unitary four-dimensionalism, and endurantism to unitary three-dimensionalism and ultimately, perdurantism is equivalent to unitary three-dimensionalism.

Now, some may find this intuitively compelling. Those who think that talk of spatially extended mereological simples that have different properties at different locations, is equivalent to talk of spatially extended mereological composites that have (spatial) parts with those properties, will have no difficulty in holding that talk of temporally extended temporal simples (terduring objects) is equivalent to talk of temporally extended objects with temporal parts (perduring objects) which is in turn equivalent to talk of persisting objects that are constituted by objects at times (enduring objects).

The crucial difference between unitary and non-unitary views hangs exclusively on the difference between an enduring fusion, at a time, and a fusion-at-a-time. While unitary theorists hold that enduring fusions constitute or compile persisting objects at times, non-unitary theorists hold that synchronic fusions constitute or are temporal parts of persisting objects at times. So both views hold that if some persisting object *O* has spatial parts *A* and *B* at t_1, then at t_1 *O* is related in some manner to a fusion of *A* and *B*. The difference is just whether this is an instantaneous fusion of *A* and *B* at t_1, or a persisting fusion of *A* and *B* *simpliciter*.

As I see it, there are two options. Those who think that there is a substantial difference between unitary and non-unitary theories, must ultimately think that there is a real difference between a persisting object being related in some way to a persisting fusion at a time, and a persisting object being related in some way to a synchronic fusion at a time. And if that is true, then the definition of constitution (and compilation) that the unitary theorist employs, ought to be relevantly different to the definition of constitution (and temporal parthood) used by the non-unitary theorist. For being related to some persisting fusion at a time, is simply not the same as being related to some synchronic fusion at that time. Then it follows that 'constitution' in the mouth of the non-unitary theorist means something different to 'constitution' in the mouth of the unitary theorist, and the transitivity argument no longer holds: we are talking of two separate relations, both of which, unhappily, have been called 'constitution'.

It might be objected, however, that we should not infer from the fact that there is a difference between a persisting object being related in some way to a persisting fusion at a time and a persisting object being related in some way to a synchronic fusion at a time, to the conclusion that the relation is different in each case. After all, why should this not simply be a case where we have the same relation—constitution (in the case of the three-dimensionalist)—which holds between different relata. For the unitary theorist this relation holds between a persisting object and a persisting fusion, and for the non-unitary theorist between a persisting object and a synchronic fusion. We would thus have a case of the same relation, but with a different relata.

But that can't be quite right. For it is not that there exist both fusions and persisting objects that materially coincide, and then we determine how these objects are related. Rather, we know that fusions are ontologically more basic, and persisting objects are ontologically dependent on these fusions for there existence: what it is for such persisting objects to exist at a time, is to be constituted by a fusion at that time. The constitution relation, recall, is an entailment relation. So it is not simply the relation that holds between two independent relata: fusion and persisting object.

Indeed, we can see how that difference might play out. Suppose the non-unitary theorist holds that some synchronic fusion F entails the existence of persisting object O at t. Since F is instantaneous, it is the existence of F *simpliciter* that entails the existence of O at t, that is, it is the existence of 'all of' F that entails the existence of O at t. Now consider some persisting fusion F^* that entails that existence of O^* at t. We can say that F^* entails O^* at t. In reality though, what is pertinent to F^* entailing O^* at t, is F^* at t. What F^* does or does not do at times other than t is irrelevant to its entailing O^* at t. So it is not that F^* *simpliciter* entails O^* at t, but rather, that F^* at t entails O^* at t. Thus one might argue that the relation between F and O is constitution, while the relation between F^* and O^* is a slightly different relation, perhaps constitution$^+$. Then it does not follow from the fact that unitary three- and four-dimensionalism are equivalent and that non-unitary three- and four-dimensionalism are equivalent, that unitary and non-unitary three- and four-dimensionalism are equivalent.

On the other hand, it is easy to see why one might embrace the idea that these views are equivalent. It is a short step indeed from considering some persisting fusion F^* at t, and holding that at t, F^* constitutes, (or constitutes$^+$) some persisting object O^* at t, to holding that F^* at t just is F. That is, it is a short step to hold that talk of a fusion, at a time, is equivalent to talk of a fusion-at-a-time. In that case, there is no difference between talk of some persisting object O being constituted at t by persisting fusion F^* at t, and talking of O being constituted by F^*-at-t, where F^*-at-t just is the synchronic fusion of the parts of F^* that exist at t. So if talk of fusions at times, really is just talk of synchronic fusions, then it follows that unitary theories will be equivalent to analogous non-unitary theories.

4.2 Unitary or Non-unitary?

So perhaps non-unitary and unitary theories really are equivalent. Then the only question at issue would be whether one ought to prefer a restricted or unrestricted theory. In what follows, however, I will assume that these views are not equivalent. Then which kind of view ought one prefer? As I see it, the more plausible accounts of persistence are the non-unitary theories. There are a number of reasons for this.

First, non-unitary theories are less 'metaphysically restricted' when it comes to the nature of simples. So long as we accept the notion of a fusion-at-a-time, the non-unitary theorist can embrace any account of simples.

Second, the non-unitary theorist can straightforwardly embrace a genuinely unrestricted composition. The unitary theorist, on the other hand, has to restrict composition to temporally overlapping objects if she wants to rule out the existence of perduring objects.[10] So talk of unrestricted composition with respect to unitary theories, is always talk of 'pseudo' unrestricted composition: unrestricted relative to temporally overlapping objects. Third, non-unitary theories have no need to be committed to the existence of two different kinds of persisting object. They do not hold that there exist both persisting fusions and non-fusions. Fourth, it seems plausible that non-unitary theories have a neater account of fission than do unitary theories. Fifth, it is arguably the case that the non-unitary account of the instantiation of temporary intrinsics is preferable. This might be debatable: it all depends on what one's intuitions are regarding intrinsic properties and their instantiation. But explicated in the way that I do, it seems that the non-unitary theorist has the best of both worlds. She gets to say that the whole persisting object has certain intrinsic properties: it has the in certain temporal ways—*t*ly or at *t*. But it really is the whole object that has those properties. And there is also a sense in which there is an object that straightforwardly has properties without any temporal modifier—the relevant synchronic fusions have properties *simpliciter*, and persisting object instantiate temporary properties in certain temporal ways, in virtue of being related to those synchronic fusions at times.

Finally, the non-unitary view seems better able to accommodate a restricted view of composition. I take it that in general, those who endorse restricted composition think that there is some plausible, non ad-hoc restriction that can be placed on composition, such that the only composite objects that exist are all and only those that feature in our folk ontology.[11] I put aside any general worries one might have about whether any such restriction could fulfil all of these desiderata. Suppose it could. Then it is easy to see how a restricted non-unitary theory might go. Whatever the relevant restriction on composition is, it will rule out the existence of various, say, spatially scattered synchronic fusions. Indeed, it can rule out the existence of all synchronic fusions except those that are related at times, to the objects that feature in our folk ontology—call them the folk objects.[12] Thus a restricted non-unitary view can say the following: the only persisting composite objects that exist are those that feature in our folk ontology, and the only instantaneous objects that exist are those that are related, at times, to those persisting objects. So there exist no massively gerrymandered spatially or temporally scattered objects. But the restricted unitarist cannot say this. For suppose that composition is restricted such that the persisting non-fusions are all and only the objects featuring in our folk ontology. Then there must also exist a range of persisting fusions: all of the fusions that are related at times, to these non-fusions. These fusions, however, are quite likely to be spatially scattered for large portions of their existence, and to be quite unlike the objects of our folk ontology. So the restricted unitarist must be committed to the existence of a range of objects that do not feature in our folk ontology. Indeed, the problem is rather greater than that. For surely any restriction on composition that rules out the existence of all non-fusions but the folk objects, will in addition rule out the existence of the very persisting fusions that constitute or compile the folk non-fusions at times. So the unitary theorist faces a dilemma: either her restriction on composition rules out the existence of various scattered and gerrymandered non-fusions, in which case it would also seem to rule out the existence of the very fusions needed to entail the existence of the non-fusions that are not ruled out, or the restriction is sufficiently weak as to allow the existence of the fusions that are necessary to entail the existence of the ordinary folk objects, in which case it seems that the restriction will also allow the existence of a plethora of gerrymandered non-fusions.

Ultimately then, these are all reasons to prefer a non-unitary theory. Moreover, notice something here. One reason we might have had to prefer a unitary theory is the intuition that where we see a single ordinary persisting object, there really does exist a single persisting object and not, as the non-unitary theorist would have it, a bunch of coinciding objects that mysteriously come into and pass out of existence at each moment. Given this rather counterintuitive consequence of non-unitary views, we might have thought that this is at least one reason to prefer unitary views. But in fact, it turns out that the unitary theorist is committed to a claim that is probably just as counterintuitive. She certainly rejects the claim that the objects that coincide with ordinary objects at times, come into and pass out of existence at those moments: she denies that they are instantaneous. But she does not deny that there are objects that coincide with ordinary objects at times: the persisting fusions that are related to those objects at times. And those fusions are no ordinary objects: they are peculiar long-lived objects that 'coalesce' at certain times, and at those time constitute or compile non-fusions. And it does not seem that *this* is any more intuitive than the non-unitary view.

If you buy all this, then what it tells us is not four-dimensionalism is preferable to three-dimensionalism; in fact it doesn't tell us anything about what we should think about persistence. Rather, what it tells us is something about a particular theoretical apparatus: it tells us that the synchronic fusion is an important theoretical and explanatory tool. Theories that appeal to the existence of such objects are much more robust, have greater explanatory resources, and are generally simpler and neater. Once we try to deny that there exist these synchronic fusions, the theories that we develop are radically different, much more elaborate, unwieldy, and generally less appealing. In that sense then, we have discovered that the synchronic fusion is *resilient* across the most plausible theories of persistence.

So where does this leave us? Well hopefully it leaves us with a clear idea of the relationship between various metaphysical theses. That is, it tells us what sorts of packages of metaphysical commitments are consistent, and which are not. It tells us that if you have certain views about composition, then you have to embrace certain views about persistence, and about the nature of simples. In general, it tells us what sorts of views one can consistently have about objects both at and across time. Finally, it tells us that for any bundle of metaphysical views paired with a three-dimensionalist account of persistence, that same bundle paired with a four-dimensionalist account of persistence will be equivalent. That is, analogous versions of three- and four-dimensionalism are metaphysically equivalent.

So the hard work is done: we have explicated the fundamentals of each theory, and shown that there is good reason to think that analogous versions are equivalent. In the next two chapters I consider two further challenges that the claim of equivalence faces. The first of these is the possibility of local backwards time travel. Given this possibility, the question arises of what we should say about a circumstance in which some time travelling object meets its younger self. I consider how each of our theories deals with this possibility, and whether, as has been claimed, the four-dimensionalist has a better account of such a situation.

NOTES

[1] I use the M-*simpliciter* locution here simply for consistency with the endurantist definition. Since in the case of synchronic fusions the M- and S-*simpliciter* notions do not come apart, either of these notions is equivalent to the straight four-dimensionalist notion of a univocal *simpliciter*.

[2] Where the 'endurantist sense' means in the language of endurantism, or by the lights of the endurantist, and the 'perdurantist sense' means in the language of perdurantism, or by the lights of the perdurantist.

[3] Lewis, (1986). pp 202–206. For arguments against the temporal relativisation of properties see also Merricks (1995) and (1994).

[4] One might suggest that the endurantist need not think that this is the direction of the dependency: she might think that it is the enduring object that is red at t_1, and the fusion is red at that time in virtue of being related to the object at that time. Maybe there is some version of endurantism that would allow you to say this, but the endurantism I have described cannot, I think, coherently say this. After all, on this account it is fusions that are ontologically basic: enduring objects get to have parts at times in virtue of being constituted by those fusions at those times. Enduring objects exist at times, in virtue of the existence of the fusion at that

time. So it only makes sense to think that they have properties at times in virtue of how they are related to fusions at times. But I don't see this as any great disadvantage.

[5] Robinson (1982).

[6] Robinson (1985).

[7] Van Inwagen (1990a).

[8] Howard-Snyder (1991).

[9] Van Inwagen (1990a).

[10] Notice there might be a complicated way around this problem. The unitary theorist could say something like this: we can only fuse temporally overlapping simples, hence there do not exist any fusions that count as perduring, and which have maximal temporal parts: there exist only 'unitary' fusions. But if composition is truly unrestricted, then there must be some objects that are in some sense 'made up of' simples that are temporally non-overlapping. So we might say that these punctuate objects exist, but they are not fusions of the temporally non-overlapping simples. Rather, they are related to those simples in some other non-mereological way. But such an account is very messy, and not at all appealing.

[11] Where 'folk ontology' is intended to include objects posited by our best science, not just those recognised by the folk.

[12] Perhaps there exists both a synchronic restriction, which might include something about the way the parts of the synchronic fusion need to be arranged, and a diachronic restriction, which might include something about a certain causal relation needing to obtain between the synchronic fusions in question.

Chapter 7

TRAVELLING IN TIME

1. THE PROBLEM OF TIME TRAVEL

The possibility of travelling back in time to a period in which one's earlier self or one's ancestors exist, raises a number of well-worn problems.[1] With respect to theories of persistence, the possibility of travelling back to a time when one's earlier self exists, raises the question of how it can be that one and the same object can exist at two different spatial locations at the same time. Call this the time traveller puzzle.

In this chapter I consider whether this puzzle can adequately be resolved by any of the theories of persistence that we have encountered, and if it can, whether one of the theories provides a better resolution than its competitors. In particular, I address the issue of whether, as Ted Sider[2] maintains, the time traveller puzzle presents a greater problem for the three-dimensionalist than for the four-dimensionalist. According to Sider, the four-dimensionalist is better able to explicate how it can be that time travellers can meet their younger selves. If this argument succeeds, then it follows that three- and four-dimensionalism have different explanatory resources. That is a problem for my account regardless of whether Sider is right to think that time travel is logically possible. For if time travel is possible, then the four-dimensionalist has greater explanatory resources. If time travel is impossible, then arguably three-dimensionalism has greater explanatory resources, since the impossibility of time travel can be inferred from the inherent difficulty (or impossibility) of explaining how one and the same object can wholly exist at multiple spatial locations at the same time. Either way, this is bad news for the claim that three- and four-dimensionalism are metaphysically equivalent.

Throughout this chapter I assume that time travel is logically possible, and attempt to show that analogous versions of three- and four-dimensionalism are equally able to cope with the time traveller puzzle. But the logical possibility of time travel is not at issue here—perhaps time travel is not logically possible. What is at issue is whether three- and four-dimensionalist theories have the same explanatory resources, and I hope that in what follows I show that they do.

I begin in section one by outlining the manner in which both 'traditional' three- and four-dimensionalists deal with the possibility of time travel. In section two I return to our earlier distinctions. I consider how unitary versions of each theory will fare in understanding how an object can travel to meet its earlier self, before in section three moving on to consider the time traveller puzzle within the context of non-unitary versions of each theory. Then in section four I consider a number of objections made by Sider, to the effect that three-dimensionalists are unable adequately to cope with the time traveller puzzle. I argue that once we see these objections in the light of the distinction between unitary and non-unitary versions of three-dimensionalism, we see that they have no force in moving anyone to prefer four-dimensionalism to three-dimensionalism. Finally, in section five I return to the puzzle of fission, and examine how the lessons learned from the time traveller puzzle might be put to use in analysing fission. Over all, I argue that consideration of the time traveller puzzle and further examination of the fission puzzle reinforces the conclusion of the previous chapter, to wit that analogous versions of three- and four-dimensionalism are equally explanatory and ultimately metaphysically equivalent.

1.1 Time Travelling Perdurantists

Suppose that an elderly Mary unearths a time machine and travels back to a time in which her younger self exists. Let us focus our attention on one pertinent temporal interval—*T*—during which elderly time travelling Mary meets the young Mary. Let 'Mary' refer to the person who is born, grows older, and eventually unearths a time machine and travels back in time. Now let us introduce some terminology that is neutral between a three- and four-dimensionalist analysis of time travel. Let us say that a spatio-temporal region is 'person-suitable' just if the distribution of intrinsic properties across that region is suitable for that region being occupied by a person during that time. During *T* there exist two relevant person-suitable regions. One of those regions is occupied by Mary when she is young. Call the occupant of that region *YS*. The other region is occupied by Mary when she is old and has travelled back in time. Call the occupant of that region *TT*.

The puzzle of time travel is what to say about the relation between *TT* and *YS*. Perdurantists are taken to have an easy solution to this problem

For the perdurantist, distinct regions of space-time that are person-suitable, contain distinct objects: in most cases they contain distinct person-stages.[3] The question then becomes whether any two or more person-stages are stages of the same persisting person. Usually when those person-stages exist at the same time the answer to that question will be no. In the case of time travel, however, it might be yes.

For the perdurantist *TT* and *YS* are distinct objects—person-stages—and they are person-stages of one and the same person: Mary. Since person-stages are temporal parts of persons, it follows that *TT* and *YS* are temporal parts of Mary. But we need to be a bit careful here. In chapter three I distinguished maximal and non-maximal temporal parts. Throughout, when talking of temporal parts I have referred to maximal temporal parts. But it is not the case that there exist multiple *maximal* temporal parts of Mary at one and the same time.

What is the relation between persons, person-stages and temporal parts? In general we want to say that persons are those four-dimensional objects that are maximally person-related, that is, they are the maximal objects that preserve personhood. Normally we would expect this to amount to a preservation of the person-relation across temporally contiguous maximal temporal parts of persons. So usually, extended maximal temporal parts and person-stages are identical. To be a stage of a person just is to be an extended maximal temporal part of a maximally person-related object.

In cases of time travel these notions come apart. When the time traveller and the young self meet, there exists one person at that time: Mary. This is reflected in the fact that the maximal temporal part of Mary at that time includes all of the spatial parts of both the time traveller and the young self. But there exist two person-stages at that time, the time travelling stage (*TT*) and the young self stage (*YS*). It is that there exist two distinct person-stages that accounts for the fact that the time traveller and the young self are distinct objects that have distinct properties. In cases of time travel then, person-stages turn out to be particular *non-maximal* temporal parts of persons.[4]

Hence Mary can meet her younger self in the past, since her younger self (*YS*) and her older self (*TT*) are distinct non-maximal temporal parts of one and the same four-dimensional whole that is Mary. Moreover, since the time travelling self and the younger self are, in the relevant sense, both Mary, we may say that during *T* Mary is both young and old. Mary has these apparently contradictory properties during *T* in virtue of there being a maximal temporal part during *T*, which has spatial parts some of which are old and some of which are young. That object is both young and old. There is also a sense in which Mary, qua time traveller, is old and not young, and Mary qua young self, is young and not old. This is the sense in which during *T* there is some non-maximal part of Mary that is young and only young: the non-maximal temporal part that overlaps all and only the young spatial parts of Mary during *T*—*YS*. There also exists, during *T*, a non-maximal part of Mary that is old and only old: *TT*.[5]

1.2 Time Travelling Three-dimensionalists

Suppose that Mary endures, and that our world is correctly described by the four-dimensional geometry of Minkowski space-time. Then the path through space-time of any persisting object is represented by that object's worldline. For the four-dimensionalist, each point on the worldline represents a three-dimensional slice of space-time that is occupied by an instantaneous temporal part of a four-dimensional object, and the worldline as a whole represents a four-dimensional volume of space-time that is occupied by the entire four-dimensional space-time worm. The three-dimensionalist agrees that any persisting object occupies a four-dimensional volume of space-time. She merely insists that each three-dimensional slice of that volume is occupied by a wholly present three-dimensional object, not an instantaneous temporal part of an object.

Regardless of whether enduring objects ever do or could travel back in time, they are wholly present at multiple regions in space-time, and have different properties at each of those regions. The difference is that in a case of time travel, the two space-time regions in question that are occupied by the time traveller and the younger self, are, from a certain frame of reference, space-like separated:[6] they exist simultaneously. In the case we are considering, the two regions of space-time in question are the two person-suitable regions of space-time, one occupied by *TT*, and the other by *YS*. What are *TT* and *YS* according to the three-dimensionalist? They are not distinct person-stages of Mary. Rather, *TT* and *YS* are simply Mary under different descriptions. Just as absent a case of time travel we can talk of young Mary and old Mary, or Mary at t_1 and Mary at t_2, talk of *TT* and *YS* is really just talk of time travelling Mary and young Mary. And just as (enduring) Mary at t_1 is identical to Mary at t_2, so too old Mary (*TT*) is identical to young Mary (*YS*). This raises the question of how Mary can have the apparently contradictory properties during *T*, of being both old and young.

But even if Mary had not travelled in time, the properties she has at one time, are different to the properties she has at another time. This problem of temporary intrinsics and the three-dimensionalist strategies for dealing with it—indexicalism and adverbialism—have already been considered at length. Recall that adverbialists deal with enduring objects being multiply *temporally* located, by holding that properties are instantiated in different *temporally* modified ways. This suggests that the three-dimensionalist can adopt an analogous strategy for dealing with enduring objects being multiply *spatially* located in cases of time travel. It suggests that in addition to relativising the having of properties to times, three-dimensionalists will also need to relativise the having of properties, to spatial regions. Then according to the spatial analog of temporal adverbialism—properties are instantiated in different spatial locational ways, or instantiated *S*ly, where *S* refers to some spatial region within which an enduring object is wholly present.

Consider a particular instant, t_1, during which Mary's time travelling self and younger self meet. Suppose that at t_1, the time travelling self (*TT*), occupies spatial region *S*, and the younger self (*YS*) occupies spatial region *S**. Suppose that at t_1, *TT* is standing and *YS* is sitting. Then at t_1 Mary has the contradictory properties of both sitting and standing. The three-dimensionalist will explain how this possible, by holding that since the time travelling self is standing at *S*, and the younger self is sitting at *S**, Mary has the properties of standing *S*ly, and sitting *S**ly.[7]

A brief aside. Since Mary has both spatially and temporally adverbialised properties, rather than talking of her having the property of standing *S*ly $t_1$ly, we could instead talk of her having that property at a space-time region. If we suppose that *TT* occupies space-time region *R*, and *YS* occupies space-time region *R**, then we can simplify matters by holding that Mary has the properties of standing *R*ly, and sitting *R**ly. The three-dimensionalist could then hold that for any space-time region *R* that is occupied by a wholly present object *O*, and any property *P* that *O* has at *R*, *O* has *P R*ly. Thus regardless of whether Mary travels in time or not, the manner of instantiation of her properties will be relativised to space-time regions. The only difference in the time travel case is that the space-time regions *R* and *R** are, from the perspective of their rest frames, space-like separated, while other pairs of space-time regions that are occupied by wholly present Mary are time-like separated.

Relativising the manner of instantiation of properties to space-time regions is the same as relativising the manner of instantiation of properties to both times and spatial locations. Frequently though, our pre-relativistic ways of thinking and speaking mean that we prefer to talk of Mary being in two different places at the same time, rather than being in two different space-time regions. Henceforth then, I will mainly talk of spatially adverbialised properties, rather than space-time regionally adverbialised properties (properties instantiated *R*ly).

But what does it really mean to have the property of sitting in an *S*ly manner? Is the notion of instantiating properties in spatially adverbialised ways a good one? The answer to this second question, according to Ted Sider, is 'no'.[8] According to Sider, the introduction of irreducibly spatially adverbialised properties is problematic on a number of fronts. I will consider these objections in section four. First, however, the former question brings us nicely to the distinction we have drawn between unitary and non-unitary versions of three- (and four-) dimensionalism. For the purposes of this chapter I will take it that *prima facie,* the distinction between unitary and non-unitary views stands: that is, I will assume that they are not metaphysically equivalent. Thus in the following section I consider how unitary versions of three- and four-dimensionalism will approach the time traveller puzzle, and I consider how to make sense of the spatial adverbialisation of properties within this context. I then move on, in section three, to consider the time traveller puzzle in the light of non-unitary versions of each theory.

2. UNITARY VIEWS

In chapter five I argued that unitary four-dimensionalists must embrace a typically three-dimensionalist account of property instantiation: they must hold that there is some instantiation relation, and I recommended temporal adverbialism. For both unitary three- and four-dimensionalists, there is some irreducible notion of instantiating a property in a particular temporally modified manner. Both persisting fusions and persisting non-fusions instantiate temporally adverbialised properties (the latter instantiate such properties because the former do). So we should expect both unitary three- and four-dimensionalists to have need of appeal to spatially adverbialised properties. They do. For the unitary four-dimensionalist cannot appeal to the existence of any non-maximal temporal parts to explain how it is that Mary has one set of properties qua time traveller and a different set qua young self. Nor can he appeal to any maximal temporal parts to explain how Mary has both the contradictory properties of sitting and standing at t_1. So the unitary four-dimensionalist must appeal to the same spatially adverbialised properties as does the (unitary) three-dimensionalist: he must say that Mary has the properties, at t_1, of standing *S*ly and sitting *S**ly.

Then the notion of having a property *P* *S*ly is *not* the notion of having some proper part at *S* that is *P*: it is not the notion of having some proper *spatial* part at S that is P, nor the notion of having some proper *temporal* part at S that is P. Rather, having a property P Sly requires that there be some *irreducible* notion of having a property in a particular spatial locational manner.

To simplify matters, let us put aside time travelling Mary for a moment, and consider some persisting fusion (enduring or terduring) O_2. Suppose O_2 is the fusion of persisting simples P, Q and, R. Suppose that O_2 travels back in time to meet itself at t_2. Suppose that at t_1, O_2 at location S_1 is the non-traveller, and O_2 at location S_2 is the time traveller. At $S_1 O_2$ is red, and at $S_2 O_2$ is blue. (Suppose the colour of O_2 at a time (and place) depend on the manner in which P, Q, and R are arranged.)

According to the unitary four-dimensionalist, one and the same terduring object—O_2—exists at S_1 and S_2. That O_2 has apparently contradictory properties at t_2 is explained by noting that it is red S_1ly and blue S_2ly at t_2. The same is true for the three-dimensionalist analysis, though she conceptualises O_2 as enduring rather than terduring. So for any persisting fusion, terduring or enduring, that travels through time to meet its earlier self, that fusion instantiates irreducibly spatially adverbialised properties just as it instantiates irreducibly temporally adverbialised properties.

Of course, for the unitary theorist Mary, and most other everyday persisting objects, are not persisting fusions: they are persisting non-fusions that are constituted by (in the case of three-dimensionalism) or compiled by (in the case of four-dimensionalism) persisting fusions at times. So let us consider some persisting non-fusion—call it O_3—which travels back in time to t_3 where it meets its earlier self. The younger self of O_3 exists at location S_1 at t_3, while the time travelling self exists at location S_2 at t_3. Suppose that O_3 is red at S_1, and blue at S_2. Once again then, the unitary theorist, whether of three- or four-dimensionalist variety, will talk of O_3 being red S_1ly at t_3, (S_1ly $t_3$ly), and being blue S_2ly at t_3 (S_2ly $t_3$ly). So do non-fusions instantiate irreducibly spatially adverbialised properties? Well non-fusions instantiate irreducibly temporally adverbialised properties, because they inherit those properties from the fusions to which they are related at times. The same is not true, however, for properties that are instantiated Sly.

To illustrate this, suppose that at S_1, O_3 is related in some manner (compiled or constituted by) persisting fusion F_1, and at S_2 is related in some manner to persisting fusion F_2. Just as any persisting non-fusion may be constituted or compiled by *distinct* persisting fusions at different temporal locations,[9] so too when there is time travel, O_3 will be constituted or compiled by distinct persisting fusions at different spatial locations. Persisting fusions F_1 and F_2 have the properties of being respectively red $t_3$ly and blue $t_3$ly: but unless those fusions are also time travellers (and let us suppose they are not) then they do not have those properties *Sly*. F_1 is red $t_3$ly, it is not red S_1ly $t_3$ly. What makes it the case that at $t_3 O_3$ is red S_1ly is that at $t_3 O_3$ is constituted or compiled, at S_3, by a persisting fusion that is red $t_3$ly. So where what constitutes or compiles persisting non-fusions at times are persisting fusions that have not themselves travelled back in time, the instantiation of properties *S*ly by those persisting non-fusions is not the instantiation of *irreducibly* spatially adverbialised properties. For what it is for some non-fusion to have a property *S*ly, is for that object to be constituted or compiled at that location, by some fusion that has that property (albeit in a temporally modified manner).

Both unitary three- and four-dimensionalists are committed to the existence of *some* irreducibly spatially adverbialised properties: they are committed to them in cases where persisting fusions travel back in time. They are also committed to the existence of reducible spatially adverbialised properties in cases where persisting non-fusions travel back in time and are constituted or compiled by multiple persisting (non-travelling) fusions at the same time. (Notice that in cases of time travel, it is important that we have multiple instances of the constitution or compilation relation holding at a time. In the case of O_3, for instance, there is some persisting fusion—call it F_3—that is the fusion of F_1 and F_2. But O_3 is not constituted or compiled at t_3, by F_3. Rather, although F_3 exists at t_3 (at least given mereological universalism), the only relation that F_3 has to O_3, is that F_3 has proper parts (F_1 and F_2) such that each of those parts constitutes or compiles O_3 at a location. For if O_3 were constituted or compiled by F_3 then (if we suppose that O_3 is a person), it would turn out that O_3 is an odd spatially scattered object that has four arms and two noses, rather than its being the case that one person with two arms and one nose, exists at multiple spatial locations at t_3.)

Where does this leave us? I noted earlier that the idea of irreducibly spatially adverbialised properties has been criticised by Ted Sider. Even if this criticism proves to be founded, it does not provide reason to think that unitary four-dimensionalism is explanatorily preferable to unitary three-dimensionalism. For both unitary three- and four-dimensionalists are committed to the existence of irreducibly spatially adverbialised properties. In fact, unitary three- and four-dimensionalism are explanatorily equivalent with respect to the time traveller puzzle. Given the suggested translation that I developed in previous chapters, it should be clear how the unitary three-dimensionalist's talk of time travel can be mapped onto the unitary four-dimensionalist's talk.

It might still be the case, of course, that Sider's criticisms provide reason to prefer non-unitary versions of three- or four-dimensionalism over unitary views, or, that they show that non-unitary four-dimensionalism—

perdurantism—has greater explanatory power than non-unitary three-dimensionalism—endurantism. I will consider these possibilities in section four. First, however, I turn to explicate the non-unitary theorist's analysis of the time traveller puzzle.

3. NON-UNITARY VIEWS

Recall that the perdurantist holds that during *T*, Mary has two non-maximal temporal parts—two person-stages—*TT* and *YS*. Let us stipulate that *TT* and *YS* are the objects that exist during and *only* during *T*. Time travelling Mary and young Mary have distinct properties because they are distinct person-stages; they are both Mary because they are both person-stages of one and the same maximally person-related object. Now consider instant t_1 that falls within *T*. Call one synchronic fusion that exists at t_1 and has the synchronic identity conditions of a person at a time, *F*. Call the other synchronic fusion that exists at t_1 and has the synchronic identity conditions of a person at a time, *F**. *F* exists at region *S*, and *F** at region *S**. Then for the perdurantist, *F* and *F** are instantaneous (non-maximal) temporal parts of Mary; *F* is a maximal instantaneous temporal part of *TT*, and *F** is a maximal instantaneous temporal part of *YS*. (*F* is *TT*-at-t_1, while *F** is *YS*-at-t_1). There also exists a fusion of *TT* and *YS*. Call that object *EF*. *EF* is an extended maximal temporal part of Mary. Further, there exists a fusion of *F* and *F**. Call that fusion *IF*. *IF* is an instantaneous maximal temporal part of Mary and of *EF*. Then it is the fact that *F* and *F** are linked by a chain of temporally contiguous synchronic fusions each of which preserves some diachronic person-identity conditions, that makes it the case that they are both temporal parts of a persisting person—Mary.

Notice that the unrestricted endurantist agrees with most of this story. She agrees that there exist two synchronic fusions at t_1, (*F* and *F**) each of which occupies a person-suitable region of space-time and has the synchronic identity conditions of a person. She agrees that there exists some fusion of *F* and *F** (*IF*). She agrees that there exists, during and only during *T*, some persisting objects that occupy person-suitable regions during *T*. Now, earlier in this chapter I noted that for the three-dimensionalist, talk of *TT* and *YS* is really just talk of Mary under different descriptions: it is talk of time travelling Mary and young self Mary. Given unrestricted versions of either unitary or non-unitary three-dimensionalism, however, this need not be the case. It is of course true that time travelling Mary and young self Mary are both Mary at two different spatio-temporal locations. So both unitary and non-unitary three-dimensionalists could use the locutions

'*TT*' and '*YS*' to refer to Mary at each of these locations. But they need not. Both unitary and non-unitary three-dimensionalists could use '*TT*' and '*YS*' to refer to relatively short-lived enduring objects that exist during and only during T, and which during that time coincide with Mary at certain locations (assuming the restricted unitary theorist thinks such objects exist). Specifically, the endurantist can hold that *TT* is the enduring object that exists only during T and which at t_1 is constituted by F, while *YS* exists only during T and at t_1 is constituted by F^*. Finally, the endurantist agrees that there exists some odd gerrymandered fusion of *TT* and *YS*, a spatially scattered object with four arms and two noses: call that fusion *EF*.

According to the endurantist, all enduring objects are at each time at which they exist, related to (constituted by) distinct synchronic fusions at each of those times. Absent cases of time travel, an enduring person is always constituted at a time, by a single synchronic fusion with the synchronic identity conditions of a person. In the absence of time travel, Mary would have been constituted by F^* and only F^* at t_1. Time travel, however, introduces the prospect of there existing multiple simultaneous synchronic fusions each of which constitutes one and the same person at a time.

So the endurantist will, like the perdurantist, hold that fusions F and F^* are distinct and have distinct properties, as are the enduring objects *TT* and *YS*. What explains why Mary has different properties qua time traveller and young self, is that qua time traveller, at t_1 she is constituted by fusion F, and qua young self, at t_1 she is constituted by fusion F^*. Nevertheless, F and F^* constitute *one and the same enduring object* at t_1: Mary. So there exists only one person at t_1, it is just that she is multiply spatially located at that time, in virtue of being constituted at that time by two distinct synchronic fusions. Thus we can explain the sense in which time travelling Mary is straightforwardly old at t_1, for at $t_1 F$ has the property of being old *simpliciter*, and Mary is constituted at S by F. So too the sense in which young self Mary is young at t_1 is explained by the fact that at S^* at t_1, Mary is constituted by F^* and F^* is young *simpliciter*. We can also explain why at t_1, Mary is both old and young; for at t_1 Mary is constituted by a fusion that is old, and is also constituted by a fusion that is young and thus Mary is indeed both old and young at t_1.

How can Mary have these contradictory properties at t_1? Because she has these properties in different spatially adverbialised ways. Mary is old and young at t_1 in virtue of being constituted, at different spatial locations S and S^*, by distinct fusions F and F^* that are, respectively, old and young *simpliciter*. Hence at t_1 Mary has the properties of being old *S*ly and young *S**ly. These spatially adverbialised properties, however, are *not* irreducible. Just as what it is for Mary to have the temporally adverbialised property of being red $t_2$ly, is for her to be constituted, at t_2, by some synchronic fusion that is red *simpliciter*, so too what it is for Mary to have the spatially adverbialised property of being old *S*ly, is for Mary to be constituted at *S*, by some synchronic fusion that has the property of being old *simpliciter.* So Mary is old *S*ly $t_1$ly, just if she is constituted at time t_1 and location *S*, by some synchronic fusion that has the property of being old *simpliciter.*

Given the framework we set up in the previous chapter which revealed how to translate endurantist talk about properties into perdurantist talk about properties, it should be straightforward to see how the translation will occur in these cases. Endurantist talk of Mary having a property *P* *S*ly is made true in virtue of Mary being appropriately related (by constitution) to some fusion (*F*) at *S*, that has *P* *simpliciter.* Just as Mary is never, say, red *simpliciter,* so too she is never old *simpliciter*, but rather, is old at a time and place, in virtue of being related at that time and place, to an object that is old *simpliciter.* The same is true for the perdurantist. Perduring Mary is not old *simpliciter*, rather, she is related in some manner at a time and place, to some object (*F*) that is old *simpliciter.* So just as the perdurantist could adopt the '*t*ly' locution as a way of talking about the properties that Mary tenselessly has, so too she could adopt the *S*ly locution as a way of talking about the properties that Mary 'spacelessly' has. For what it is for Mary to be old *S*ly and young *S**ly, is, by the perdurantist's lights, for Mary to have some temporal part at *S* that is old, and a temporal part at *S** that is young.

At this point one might be tempted to point out that there is a disanalogy between the endurantist and the perdurantist analysis of time travel. I have suggested that for the perdurantist, what makes it true that Mary is both old and young at t_1, is that at t_1 there exists some maximal temporal part of Mary that has (proper spatial) parts that are young, and (proper spatial) parts that are old. This analysis is a little different from the endurantist analysis, according to which at t_1, Mary is constituted by both a fusion that is old, and a fusion that is young. In particular, in chapter six I argued that having some synchronic fusion as a temporal part at *t* (for the perdurantist) is equivalent to being constituted by some synchronic fusion at *t* (for the endurantist). I also noted that I intended 'temporal part' to be read 'maximal temporal part'. But given the possibility of time travel, that cannot be quite right.

For it is not the case that at t_1 the endurantist will hold that Mary is constituted by some synchronic fusion—call it *F*—such that the perdurantist holds that at t_1 *F* is a maximal temporal part of Mary. The perdurantist holds that *IF* is the maximal temporal part of Mary at t_1, and the endurantist does not think that *IF* constitutes Mary at t_1. The endurantist agrees that *IF* exists at t_1. She merely disagrees that *IF* constitutes Mary at t_1, or that what explains why Mary is both young and old at t_1 is the existence of *IF*.

In fact though, when we think about perdurantism a little more, we see that something similar is true here. Everyone (all unrestricted unitary theorists) agrees that *IF* exists at t_1, and is an odd spatially scattered gerrymandered object. But is it really the existence of this object that explains the sense in which Mary is both young and old at t_1? For the perdurantist, what really explains why Mary is both young and old at t_1, is that at t_1 there exist two person-stages of one and the same person—Mary—one of which is old and the other of which is young. That there also exists some gerrymandered fusion of those person-stages does not add any explanatory power to the story. Ultimately, both endurantist and perdurantist agree that Mary is both young and old at t_1, in virtue of Mary existing in two distinct locations at t_1 such that at one location she is old, and the other location she is young.

So what should the endurantist say about the relation between Mary and *IF*? *IF* does not constitute Mary, and yet *IF* is, for the perdurantist, a maximal temporal part of Mary. What is clear is that in cases where there is no time travel, having a maximal temporal part at a time, and being constituted by some fusion at a time, are equivalent notions (if the translation is a correct one as I have argued). Equally, though, this is not true when we are considering cases where time travelling objects exist at the same time as their earlier selves. The more general translation schema is one according to which the perdurantist claim that there exists at t some synchronic fusion F such that F is a maximal temporal part of O at t, is equivalent to the endurantist claim that there exists some synchronic fusion F at t, such that F has parts each of which constitute O at t, and such that those parts are an exhaustive decomposition of F.

So from the perdurantist perspective, fusion *IF* is a maximal temporal part of Mary at t_1. From the endurantist perspective, *IF* has proper parts, F and F^*, that exhaustively decompose *IF*, and such that both F and F^* constitute Mary at t_1. Notice that this makes good sense of cases of multiple time travel: cases where an individual travels back in time to the same temporal location multiple times. We can imagine Mary travelling back in time to t_1 five times, such that there exist six Mary's at t_1. Then the perdurantist will hold that the massively gerrymandered object that is the fusion of each of Mary's person-stages at t_1—call it *MF*—is Mary's maximal temporal part at t_1. The endurantist will hold that *MF* is some object that has proper parts (six of them) each of which constitute Mary at t_1, and such that these parts exhaustively decompose *MF*.

So in cases of time travel, having some fusion F as a non-maximal person-stage, and being constituted by fusion F at a time and place, turn out to be equivalent notions. Notice that just as the notion of a maximal temporal part has a hidden temporal index, so too the notion of a non-maximal temporal part has a hidden spatial index. What it is to be an instantaneous non-maximal temporal part of some persisting object *O*, is to overlap at a time, certain (proper) spatial parts of that object at that time. Thus we should really say that being related (by parthood) to some fusion *F* at a time and location is equivalent to being related (by constitution) to some fusion *F* at a time and location. In Mary's case, having fusion *F* as a (non-maximal) part at t_1 at *S*, is equivalent to being constituted by fusion *F* at t_1 at *S*. Thus in general, talk of having a property *P S*ly, is equivalent to talk of having some non-maximal part at *S*, that is *P*.

> Moreover, we can see why the more general translation between: there exists at *t* some synchronic fusion *F* such that *F* is a maximal temporal part of *O* at *t*.
>
> and
>
> there exists at *t* some synchronic fusion *F*, such that *F* has parts each of which constitute *O* at *t*, and such that those parts are an exhaustive decomposition of *F*.

encapsulates the normal cases where there is no time travel. Absent time travel, the part of *F* that constitutes *O* at *t* is an improper part: *F* itself constitutes *O* at *t*. So absent time travel, having a fusion *F* as a maximal temporal part at *t* is equivalent to being constituted at *t* by some fusion *F*.

So far then, it looks as though endurantists and perdurantists deal with the time traveller puzzle in the same way: both have the same explanatory resources, and it is straightforward to see how the translation will proceed when we introduce spatially adverbialised properties. In the following sections, however, I consider some objections levelled against the 'traditional' three-dimensionalist account of time travel that appeals to these spatially adverbialised properties. For perhaps these objections are equally telling against any view that appeals to spatially adverbialised properties.

4. AGAINST SPATIAL ADVERBIALISATION

The first of the aforementioned objections is that, according to Sider, one who employs spatially adverbialised properties in analysing time traveller puzzles is unable to distinguish between the properties an object actually has, and the properties it would have had if things had gone differently.[10] I consider this objection in the next section, before turning in section 4.2 to consider a rather different objection, namely that embracing spatially adverbialised properties might lead one to disavow the existence of any parts, that is, it might lead one to reject mereology outright. First to the former objection.

4.1 Knowing Which of Me Does What

Here's how the objection goes. The spatial adverbialist is unable to tell the difference between a world in which Mary's time travelling self stands and her younger self sits, and a world in which her time travelling self sits, and her younger self stands. We know that in the actual world—call it w—at t_1 Mary has the properties of standing Sly and sitting S^*ly in virtue of her time travelling self standing at S, and her younger self sitting at S^*. But suppose things had happened a little differently. Consider the counterfactual world w^* in which at t_1 Mary's time travelling self occupies spatial location S^* and is sitting, and her young self occupies spatial location S and is standing. In w^* Mary has the properties of sitting S^*ly and standing Sly at t_1, the very same properties she has in the actual world, despite the fact that in w it is the time travelling self who is sitting and not the young self.

The first thing to notice about this argument is that it is sound only given that having a property P Sly is having P in an *irreducibly* spatially adverbialised manner. Just as the problem Sider raises is no problem for the perdurantist, who can simply say that in w^* it is a different person-stage that is sitting than the person-stage that is sitting in w, so too the problem does not arise for the endurantist. For although the endurantist appeals to spatially adverbialised properties, she does not appeal to *irreducibly* spatially adverbialised properties. Thus the endurantist can also explain the difference between w and w^*, by noting that in w Mary has the property of standing Sly in virtue of being constituted by fusion F at S, while in w^* she has that property in virtue of being constituted by a different fusion at S, namely F^*. So long as having a property P Sly is reducible in some way—either to having some fusion F at a time/location as a (non-maximal temporal) part that has P *simpliciter*, or to being constituted at a time/location by some fusion F that has P *simpliciter*—then it will always be possible to distinguish the sorts of cases Sider is concerned with. Hence both endurantist and perdurantist are equally able to cope with this objection.

Indeed, in many cases the unitary three- and four-dimensionalist will also be able to distinguish these sorts of cases. As we saw in section 2, where persisting non-fusions are constituted or compiled by non-time travelling fusions at times, the spatially adverbialised properties of those non-fusions will not be irreducible. However, all unitary three- and four-dimensionalists are committed to the existence of *some* irreducibly spatially adverbialised properties. But if that commitment is problematic, then it is equally problematic for three- and four-dimensionalist versions of a unitary theory. In what follows I attempt to show that the objection mounted by Sider is by no means a devastating one, though the thesis that analogous versions of unitary three- and four-dimensionalism are equivalent does not rest on this being the case.

4.1.1 Irreducibly Spatially Adverbialised Properties

Let us consider Sider's objection within the context of unitary versions of three- and four-dimensionalism. It is true with respect to the properties of sitting and standing, that at t_1 Mary has those properties in the same way in w as she does in w^*. (I use the example of Mary, despite the fact that Mary is likely a persisting non-fusion, and thus so long as the fusions that constitute or compile her at S and S^* are not time travellers, the problem does not arise. For the sake of clarity and continuity however, let us suppose that Mary is a persisting fusion in order that we might better consider this objection). But it is not true that Mary will have *all* of the same properties in w and w^*, and thus not true that the unitary theorist is unable to distinguish the two situations.

To see this, suppose that at t_1 Mary's younger self has blonde hair and her time travelling self has grey hair. Then in w at t_1 Mary has the properties of having blonde hair S^*ly, and grey hair Sly. In w^*, however, Mary has the properties of having blonde hair Sly and grey hair S^*ly. In the world in which the time traveller and young self's locations are reversed, the manner in which Mary has the properties of having blonde and grey hair are also reversed. So in w, the totality of Mary's $t_1$ly properties are different to the totality of Mary's $t_1$ly properties in w^*. Indeed, in w Mary has different second-order properties than she does in w^*. In w she has the second-order property of having the property of sitting in the same spatially adverbialised manner as she has the property of being blonde: namely she has each of these properties S^*ly. In w^* she has the second-order property of having the property of sitting in the same spatially adverbialised manner as she has the property of being grey: the S^*ly manner. And that's because in the actual world, Mary has those properties in virtue of having a younger self present at S^*, whereas in the counterfactual world she has those properties in virtue of having a time travelling self present at S^*.

So as long as we assume that for any time t at which some persisting object O is present at multiple spatial locations S and S^*, there is some property P that O has at S and lacks at S^*, then it will always be possible to distinguish the sort of 'role and spatial' reversal cases that Sider is concerned with. In general we would expect this to be the case, even where we are considering persisting fusions. And even though persisting fusions that time travel have their properties in irreducibly spatially adverbialised ways, it will still be possible to distinguish world *w* from *w** in virtue of the fact that such fusions will have different *S*ly properties in *w*, from the *S*ly properties that they will have in *w**.

But what of occasions where a persisting fusion has all of the same intrinsic properties whenever it exists, and thus has the same intrinsic properties at each location when it travels back in time? Under these circumstances a time travelling object—call it *PO*—will have the same properties in the actual world *w*, as it has in *w**. But it is still true that at one of the locations at which *PO* is present, it has the property of being a time traveller, and at the other location it lacks that property. The property of being a time traveller is the relational property of having certain causal relations to future events. So our persisting fusion will have different relational properties in *w* than it does in *w**. In *w* *PO* has the property of being a time traveller *S*ly, (let us suppose) and the property of not being a time traveller *S**ly. The reverse is true in *w**. Of course, all of *PO*'s other properties are the same. If *PO* is red *S*ly in *w*, then *PO* is also red *S*ly in *w**. But then, if *PO* is red *S*ly in *w*, then it is also red *S**ly in *w*. The point about *PO* is that it has the same intrinsic properties whenever it exists. So it should hardly be surprising that when it travels back in time to meet itself, we have difficulty telling the time traveller apart from the non-traveller, since all that sets apart *PO* the traveller from *PO* the non-traveller is precisely that the former travelled back in time and the latter did not. This is exactly what is revealed by the fact that the only properties time travelling *PO* and non-travelling *PO* do not share, is the relational property of being a time traveller. And since any persisting object is only a time traveller if it has the relevant causal connections to the future, it will always be the case that such scenarios can be distinguished, albeit by reference to relations rather than properties. Perhaps this appeal to relational properties represents some minor drawback for the unitary theorist, and might therefore support my earlier contention that we should prefer a non-unitary theory. But it is not clear that this is by any means one of the stronger arguments in favour of a non-unitary view. And these considerations certainly do not show that four-dimensionalism has any explanatory advantage that three-dimensionalism lacks, or more specifically, that unitary four-dimensionalism has any advantage over unitary three-dimensionalism.

One objection down, one to go. In the following section I consider the more serious objection to spatially adverbialised properties, according to which once you let a spatially adverbialised property in your metaphysical door, you might as well usher out parthood for good. Can this really be so?

4.2 Mereological Abstinence

According to this objection, once we introduce the notion of instantiating properties in spatially adverbialised ways, the question presents itself as to why we should *ever* treat spatial extension and persistence differently. Why not reject the idea that spatially extended objects exist at different spatial locations in virtue of having spatial parts that exists at each of those locations? For if the notion of having properties *S*ly is a coherent one, this paves the way for arguing that we should treat local intrinsics in the same manner as we treat temporary intrinsics. Just as we can explain how objects both change over time and exist at multiple temporal locations by appealing to temporally adverbialised properties rather than temporal parts, so too we can explain how objects exist at multiple spatial locations and change across space, by appealing to spatially adverbialised properties rather than spatial parts.[11] Since local intrinsics can be explained without appeal to mereology, there is no need to posit the existence of spatial parts.

This argument, however, is not convincing. The argument is intended to lead one to prefer perdurantism, to 'traditional' three-dimensionalism, via the claim that given the possibility of time travel, extension through time and space should be treated analogously. Plausibly, most three-dimensionalists are more committed to explicating local intrinsics in terms of the existence of spatial parts, than they are committed to *denying* that there exist any temporal parts. So if space and time are to be treated analogously, this strongly suggests that even three-dimensionalists will ultimately prefer perdurantism to the view that there exist neither spatial nor temporal parts. Given the distinctions we have drawn throughout this book, however, it should be clear that the relevant dichotomy is not between perdurantism and traditional three-dimensionalism, where 'traditional three-dimensionalism' in this context is a view that is closest to the view I have called unitary three-dimensionalism.

As with the previous objection, this argument succeeds only given that the spatially adverbialised properties in question are irreducible. If what it is to have some property *P* *S*ly is to have some part at *S* that is *P*, then clearly the introduction of *S*ly properties does not mark the demise of mereology. This is precisely what the endurantist holds: an enduring object *O* has a property *P* *S*ly just if at *S*, *O* is constituted by some fusion *F* that is *P simpliciter*. So *O* has *P* *S*ly just if *O* has some improper part at *S*, that is *P*.

So this argument fails if we are considering non-unitary three-dimensionalism. But the unitary three- and four-dimensionalist are committed to the existence of irreducibly spatially adverbialised properties. So are these views in danger of either rejecting mereology outright, or of ultimately being forced to accept perdurantism? No. Even if it were true that accepting irreducibly spatially adverbialised properties inevitably leads to a rejection of mereology, it does not follow that those who find such an outcome repugnant must instead embrace perdurantism; they might just as easily embrace endurantism, given that both of these views are immune to the objection at hand. But is it true that letting a few spatially adverbialised properties into your ontology can be so catastrophic?

I think not. Recall that there are those who consider it at least logically possible that there exist spatially extended mereological simples that have different properties at different locations.[12] We can imagine a simple—O_1—that is half red and half yellow. Since we cannot explain the local intrinsics of O_1 in terms of its having proper spatial parts with certain properties, an obvious suggestion is to analyse O_1*'s* properties in terms of an instantiation relation that is spatially modified. Then, all too familiarly, we can say that O_1 is red at location *S*, and yellow at location *S** and thus has the properties of being red *S*ly and yellow *S**ly.

But, the argument might then proceed, if we need to introduce (irreducibly) spatially adverbialised properties to explain the local intrinsics of spatially extended mereological simples, then why not use such a strategy to explain local intrinsics in general? If such simples are possible, why not in effect maintain that all spatially extended objects are these simples. After all, this latter theory is a simpler one to a theory in which there exist spatially extended objects, some of which are mereologically simple and have local properties that are spatially adverbialised, and others of which are mereologically composite and have local properties that are properties of spatial parts.

So if the mere spectre of irreducibly spatially adverbialised properties is sufficient to threaten the metaphysics of mereology, then it is not only unitary three- and four-dimensionalists who should be worried. But of course, while it is plausible that all things being equal we should prefer the simpler theory, none of this goes any way to showing that all things are equal. The best unitary theory is surely one that reconciles the possibility of time travel, with the intuition that persisting objects are composed of spatial parts at times. This is something that both the unitary three- and four-dimensionalist agree on. Both agree that persisting through time is disanalogous to extending through space, though by their own lights, they disagree about why this is so: three-dimensionalists think it is because persisting objects endure, and four-dimensionalists think it is because they terdure. So the analogy above with spatially extended mereological simples is rather a nice one. Just as unitary four-dimensionalists think that persisting objects are temporally extended temporal simples, so too there are those who hold that there exist spatially extended 'spatial' simples. But it is not clear why holding either of these views ought lead one to a wholesale rejection of the existence of spatial parts.

Rather, it seems fairly clear that unitary theorists of both stripes need to make the perfectly principled distinction between relativising the instantiation of properties to spatial regions, and relativising them to space-time points. The latter view effectively has it that objects endure or terdure across space as well as time, and thus collapses the distinction between spatial extension and persistence in favour of a complete rejection of mereology. The former view allows that objects that travel back in time exist at multiple spatial regions, and thus that *within* each of these regions these objects have spatial parts. It is this view that both unitary three- and four-dimensionalist will embrace. Of course, this presupposes that spatially extended persisting objects are composed of spatial parts at a time. But this is precisely what the unitary three- and four-dimensionalist do presuppose. The unitary theorist only introduces *irreducibly* spatially adverbialised properties in the context of time travelling persisting fusions. But naturally, persisting *fusions* have spatial parts, otherwise they wouldn't be fusions. So the spatially adverbialised properties in question must be relativised to spatial regions and not to space-time points—that simply follows from the fact that it is persisting fusions that are the bearers of the spatially adverbialised properties. Thus the existence of such properties cannot be reason to question the existence of spatial parts in general.

With these two objections out of the way one thing should be perfectly clear: whatever the concerns one might have about irreducibly spatially adverbialised properties, they do not militate in favour of four-dimensionalism over three-dimensionalism. They may lead one to prefer non-unitary versions of three- or four-dimensionalism, but they cannot lead one to prefer unitary four-dimensionalism to unitary three-dimensionalism (or *vice versa)* or perdurantism to endurantism (or *vice versa*). Instead, when we consider the manner in which analogous views analyse the time traveller puzzle, we see further evidence for the thesis that three- and four-dimensionalism are metaphysically equivalent.

So where to now? In previous chapters I have examined how analogous versions of three- and four-dimensionalism analyse fission, and I have hinted at an analysis of fission that has hitherto not been mentioned. It is to this analysis that I turn now.

5. FISSION RE-EXAMINED

When I set up the puzzle of fission in chapter one, I blithely noted that the pre-fission Riker cannot be identical to both post-fission persons, R_1 and R_2, because identity is transitive, and R_1 is not identical to R_2. R_1 and R_2 are *qualitatively* identical, I said, but they are not *numerically* identical, since there are two of them: they exist in two distinct spatial locations at the same time.

But if time travel is possible, then the three-dimensionalist has to deny the apparent truism that numerically identical objects cannot occupy distinct spatial regions at the same time. This opens up the possibility of treating cases of fission in an analogous manner to cases of time travel. The three-dimensionalist can say that R_1 and R_2 really are both Riker: they are strictly identical. After fission, Riker simply wholly exists at multiple spatial regions. This preserves both the intuition that Riker survives fission and does so in virtue of being strictly identical with himself across time, and also that there is no principled way of declaring one of the resultant persons to be Riker, and the other not. This latter is true because they are, as intuition tells us, both Riker. Fission just is that peculiar process that causes a single individual to be multiply spatially located. Post-fission then, the manner of instantiation of all of Riker's properties will be relativised to spatial regions. Riker can be both a bachelor and married, can have children and be childless, by having each of these contradictory properties in different spatially adverbialised ways.

Post-fission then, at each time at which he exists Riker has properties in different spatially adverbialised ways in virtue of being constituted, at different spatial locations, by different fusions—enduring fusions in the case of the unitary three-dimensionalist, and synchronic fusions in the case of the non-unitary three-dimensionalist.

How does this three-dimensionalist account of fission fit in with the three- and four-dimensionalist accounts that we have previously considered? Presumably almost *any* version of unitary or non-unitary three-dimensionalism will hold that there exist objects R_1 and R_2 which temporarily coincide (coincide pre but not post-fission) According to this new analysis though, these objects are not Riker. Rather, Riker coincides with these objects at various times. Post-fission, Riker at one spatial location coincides with R_1 at times, and Riker at another other spatial location coincides with R_2 at times. That is, post-fission, R_1 and Riker are co-constituted at times and locations, as are R_2 and Riker.

So the difference between this analysis and the previous three-dimensionalist analyses that we have considered comes down to the relationship between

Riker and R_1 and R_2. The previous analyses hold that the name Riker is ambiguous between referring to each of two distinct persons—$R_1$and R_2—while the latter analysis holds that the name Riker uniquely refers, it refers to an object which post-fission is multiply located and is co-constituted with each of R_1 and R_2 at locations and times. Notice that we could think of this difference as merely semantic. All three-dimensionalists *could* concede that there exists some object that post-fission is multiply spatially located, and which at each of those locations is co-constituted with R_1 and R_2. Some might think that this object is rightly uniquely referred to as 'Riker', while others might think that this object is not Riker, but rather, that 'Riker' refers ambiguously to R_1 and R_2.

The perdurantist analog of this view is then pretty straightforward. The perdurantist will hold that there exist both R_1 and R_2 which are distinct space-time worms that share maximal temporal parts prior to fission. But on this view, Riker is not identical to either of these worms. Nor does the name 'Riker' refer ambiguously to these worms. Rather, 'Riker' refers uniquely to a space-time worm that shares its maximal temporal parts with both R_1 and R_2 prior to fission. Post-fission though, this space-time worm is spatially scattered: it is multiply located. Post-fission, R_1 and R_2 represent different person-stages of one and the same worm: the Riker worm. They are merely person-stages that exist at the same time. So R_1 and R_2 are each non-maximal temporal parts of Riker post-fission.

Similarly, the unitary four-dimensionalist will hold that R_1 and R_2 are distinct terduring non-fusions that are each compiled by distinct terduring fusions at times. But neither R_1 nor R_2 are Riker. Rather, post-fission, at each time at which he exists, Riker is simultaneously compiled by two distinct terduring fusions, one of which compiles R_1 post-fission, and the other of which compiles R_2 post-fission. Hence Riker has various spatially adverbialised properties in virtue of being compiled, at different locations, by different terduring fusions.

This analysis then, puts a different slant on the puzzle of fission, while allowing us to preserve the same intuitions as the earlier analysis.

6. WHERE TO FROM HERE?

What does all this tell us? An examination of the time traveller puzzle reveals that analogous versions of three- and four-dimensionalism are equally able to explicate how it is possible to travel back in time and meet one's earlier self. Moreover, following the thread of time travel into the puzzle of fission, we find another potential analysis of fission that reconciles our competing intuitions about such cases. This analysis is also one that can be embraced by three- and four-dimensionalists alike. Whether this analysis is to be preferred or not is an open question. Consideration of these two puzzles and their attendant solutions, however, strengthens the case for the metaphysical equivalence of three-and four-dimensionalism. For once again we find that each provides us with the same resources.

There is one issue that has yet to be tackled. I noted in chapter five that *prima facie* it appears that three- and four-dimensionalism are empirically equivalent, and I have assumed throughout that this is the case. This assumption, however, might be flawed. Perhaps scientific discovery reveals that these theories are not empirically equivalent and hence not metaphysically equivalent. It is to this issue that I turn in the last chapter.

NOTES

[1] For discussions of the grandfather paradox, or the paradox of auto-infanticide, see Grey (1999); Chambers (1999); Horwich (1975) and Sider (2002a).

[2] see Sider (2001).

[3] Except in odd cases where we have swamp persons, (objects that are qualitatively like persons at times, but which coalesce out of a swamp, fully formed) or in odd cases where we have only a very short-lived object that looks like a person-stage, but where there is no person for it to be a stage of.

[4] Not every non-maximal temporal part will of a person is a person stage.

[5] This view was originally defended in Lewis (1976).

[6] Where *x* and *y* are space-like separated just if there is no causal signal that can pass between *x* and *y*, that is, *x* and *y* are *only* space-like separated: they are simultaneous from some reference frame.

[7] This is the view that Sider countenances in his (2001) pp 102–103, and endorses in his (2002a) p 133.

[8] Sider, (2001) pp 102–104.

[9] Non-fusions might be constituted or compiled by the same fusion at more than one time.

[10] Sider, (2001) pp 102–104.

[11] Sider mentions this worry in his (2001) p 105.

[12] See for example Parsons (2000) and Markosian (1998) and (2004).

Chapter 8

EMPIRICAL EQUIVALENCE AND SPECIAL RELATIVITY

In earlier chapters I noted that *prima facie*, three- and four-dimensionalism appear to be empirically equivalent. There seems to be no empirical discovery we could make that would enable us to discern whether, at any particular temporal location, there exists some wholly present object, some temporal part of some object, or a temporally extended simple 'at a location'. That is important. Any two theories are metaphysically equivalent only if they are empirically equivalent. So if we are to provide good evidence that three- and four-dimensionalism are equivalent, it will be necessary to show that they are empirically equivalent.

Recently it has been argued that considerations drawn from empirical findings about the nature of space and time, in particular the theory of special relativity, provide reasons to prefer four- to three-dimensionalism.[1] The strength of this claim varies. At one end of the spectrum we find the claim that three-dimensionalism is straightforwardly inconsistent with special relativity, and that therefore three-dimensionalism is false. Call this *the strong view*. A more moderate view is that three-dimensionalism can be made consistent with special relativity only in an outmoded and bizarre way that requires considerable theoretical contortion, and has many counter-intuitive consequences.[2] Call this *the moderate view*. At the furthest end of the spectrum is the view that both three- and four-dimensionalism are perfectly consistent with special relativity, but that within the empirical context of four-dimensional Minkowski space-time, four-dimensionalism has explanatory resources that three-dimensionalism lacks. Call this *the weak view*.

If any of the strong, moderate, or weak views is true, then my contention that three- and four-dimensionalism are equivalent is in trouble, for even on the weak view, it turns out that there are explanatory differences between three- and four-dimensionalism. In what follows I consider each of these views. I begin in section one by considering two arguments for the strong view. The first, which proceeds via the claim that presentism is inconsistent with special relativity, I consider only briefly. The second purports to show that given a plausible account of co-existence, three-dimensionalists are committed to enduring objects being composed of non-present parts and thus are committed to enduring objects not being wholly present whenever they exist. I argue that once we clarify the three-dimensionalist's notion of co-existence, we see that this argument fails.

In section two I consider two arguments for the moderate view. The first contends that although the three-dimensionalist can construct a coherent account of co-existence within the context of special relativity, employing this notion has counterintuitive consequences. It turns out that at a single location in space-time, one object can co-exist with an enduring object multiple times. Moreover, according to the second argument, three-dimensional objects must be unacceptably relativistic in nature: they must have different parts and properties relative to different frames of reference. Since both of these claims are deeply counterintuitive, we have reason to suppose that three-dimensionalism does not accord well with our empirical findings about the nature of space-time. I argue that although for the most part three-dimensionalism does entail the consequences in question, these consequences are no more counterintuitive than the ones to which the four-dimensionalist is committed. Our intuitions are strictly pre-relativistic, and thus we find much about our relativistic universe counterintuitive. But this will be true regardless of which theory of persistence one adopts.

Then in section three I consider whether there is reason to think that four-dimensionalism has explanatory resources that three-dimensionalism lacks: is the weak view correct? I argue that it is not. Lastly, in section four I consider one final issue that might lead one to conclude that there are empirical and/or explanatory differences between three- and four-dimensionalism. I consider an issue regarding the relationship between volumes of space-time and the objects that exist 'at', or 'within' those volumes. There are three different views about the relation between objects and regions and space-time: substantival dualism, substantival monism, and relational monism. It appears that while the four-dimensionalism can embrace any of these three views, the three-dimensionalist is limited to just two: she must reject substantival monism. If that is so, then it looks like there is a genuine difference between these two theories. I argue that appearances to the contrary, no such difference exists, or at least, that given what we already know about each of these theories, there is good reason to suppose that no such difference exists.

Let us begin by considering the strong view.

1. THE STRONG VIEW

The theory of special relativity states that observers in different frames of reference will disagree about the spatial distances between objects and the durations between events. This latter is to say that there is no absolute simultaneity: while from one frame of reference R_1 two events E_1 and E_2 are simultaneous, from some other frame of reference R_2, events E_1 and E_2 are not simultaneous. It is alleged by the strong view that this empirical finding is inconsistent with three-dimensionalism. Why so?

1.1 Presentism, Three-dimensionalism and Special Relativity

One might hold that three-dimensionalism is inconsistent with special relativity, because three-dimensionalism entails presentism, and presentism is inconsistent with special relativity. The argument for this latter claim rests on the idea that special relativity is inconsistent with the notion of a single objective present wherein all and only the ontologically real things exist. Since special relativity tells us that there is no unique class of events that are simultaneous from all frames of reference, and since there is no privileged frame of reference, there is no way to determine which events are present, and hence, given presentism, which events exist.[3] At the very least, if presentism is not inconsistent with special relativity, some independent account must be given of this objective notion of 'the present' such that, perhaps, which events are truly objectively present is not a matter of which events are simultaneous from some reference frame—sometimes from some reference frame, events that are present will be simultaneous with events that are not present, and sometimes from some reference frame, present event will not be simultaneous with each other. (This might require some notion of simultaneity* which holds between all and only the present times). Or alternatively, the presentist could take a stab at arguing that there is only one privileged reference frame, and it is from the perspective of this frame that we determine which events are present.

I will not consider any of these issues. Perhaps considerations of this kind show that presentism is inconsistent with special relativity. Suppose that is true. We noted in chapters one and two that it is not uncommon to hold that presentism and three-dimensionalism are in some way deeply inter-related doctrines,[4] or that three-dimensionalism is coherent only given presentism.[5] If this latter were true, then if presentism is inconsistent with special relativity it would follow that three-dimensionalism is too. That would be a problem. I have argued throughout, however, that three-dimensionalism is *not* committed to presentism: eternalist versions of three-dimensionalism are perfectly coherent. So these types of arguments against presentism, even if successful, do not militate in favour of four-dimensionalism, they merely show that whichever theory of persistence we adopt, we should, in addition embrace an eternalist theory of time.

But there is another argument that aims to show that three-dimensionalism is inconsistent with special relativity, by showing that the idea of objects being wholly present at a time is incoherent given that there is no absolute simultaneity.

1.2 Special Relativity and Endurance

Hales and Johnson have recently argued that if we take special relativity seriously, it follows that persisting objects are composed of non-present parts.[6] If that is true, then those objects are not wholly present whenever they exist: they do not endure. So, they argue, special relativity is inconsistent with three-dimensionalism. Their argument proceeds as follows:[7]

1. Assumption: Special relativity is true.
2. Assumption: Three-dimensionalism is the thesis that objects are wholly present whenever they exist.
3. Assumption: Simultaneity is sufficient for co-existence.
4. Assumption: Co-existence is transitive.
5. An object O is wholly present at t iff all of its parts co-exist at that time.
6. Let P and Q be two enduring objects that are proper parts of enduring object O.
7. Let P_1 and Q_1 represent two points on the worldline of P and Q such that P_1 and Q_1 are simultaneous at t_1 in O's rest frame R.
8. Let P_2 and Q_2 represent two points on the worldline of P and Q such that P_2 and Q_2 are simultaneous at t_2 in O's rest frame R.
9. Let P_2 be in the absolute future of P_1, and Q_2 be in the absolute future of Q_2, where x is in the absolute future of y iff there is some causal signal which can travel from y to x.
10. So P_1 and Q_1 co-exist and P_2 and Q_2 co-exist. (3,7,8)
11. There is some frame of reference R^* from the perspective of which P_1 and Q_2 are simultaneous.
12. So P_1 co-exists with P_2 (4,10,11).
13. So all of O's parts do not co-exist at the same time (9,12).
14. Therefore O is not wholly present.

Notice assumption (4). Hales and Johnson claim that 'it is natural and common to assume that co-existence is transitive'. Given that our folk intuitions are stubbornly pre-relativistic, it is *natural* to think that co-existence is transitive. But given our pre-relativistic intuitions, it is also natural to think of simultaneity as transitive. It is tempting to say that *A* and *B* co-exist just if *A* is simultaneous with *B* (relative to some frame). It is also tempting to say that if *A* and *B* are simultaneous, and *B* and *C* are simultaneous, then *A* and *C* are simultaneous and thus *A* and *C* co-exist. Tempting though this would be, it would be false. Simultaneity is not transitive in general. Specifically, simultaneity is not transitive across different frames of reference. If *A* is simultaneous with *B* relative to frame *R*, and *B* is simultaneous with *C* relative to frame *R**, then it does not follow that there is any frame relative to which *A* is simultaneous with *C*. But if simultaneity grounds co-existence, why should we think that co-existence is transitive?

If *A* and *B* co-exist iff *A* and *B* are simultaneous relative to some reference frame, then we ought not think that co-existence is transitive. For if simultaneity is not only *sufficient* for co-existence, as Hales and Johnson assume in premise 3, but is also *necessary*, then co-existence is not transitive if simultaneity is not.

So why do Hales and Johnson think that co-existence is transitive? They cite Putnam, among others, as a source of this view.[8] But it is not the kinds of pre-relativistic intuitions we just described, that Putnam has in mind when he says that co-existence is transitive. For Putnam, the universe is a block universe in which all four-dimensional objects tenselessly exist. So *all* objects co-exist.[9] Since on this view all it takes to co-exist is to exist in space-time, it trivially follows that co-existence is transitive.

It is clear though, that this minimalist reading of co-existence according to which everything co-exists with everything else, is not what Hales and Johnson mean to capture by the term. What they mean to capture is the sense in which I co-exist with my dog in a way that I do not co-exist with Caesar. It is only this that makes it sensible to define being wholly present in terms of the co-existence of parts at a time (on the minimalist reading it makes no sense to talk of objects co-existing *at a time*).

So the three-dimensionalist *could* accept Hales' and Johnson's definition of co-existence according to which simultaneity is merely sufficient for co-existence, and could thus accept that co-existence is transitive. Then it follows that P_1 and P_2 co-exist. But if *that* is what we mean by co-existence, there seems no reason that the three-dimensionalist would accept (5), which defines being wholly present in terms of co-existence of parts at a time. The three-dimensionalist holds that an object is wholly present at a time just if all of its parts (S-*simpliciter*) exist *at that time*: just if all its parts

(S-*simpliciter*) are simultaneous at t. If we understand co-existence as Hales and Johnson do, however, it turns out that P_1 and P_2 co-exist despite the fact that P_2 is future relative to P_1. In their sense of co-existence, there is no reason to suppose that all of an enduring object's parts co-exist at the same time. For to say that P_1 and P_2 co-exist in this sense, is just to say that P_2 is simultaneous with Q_2 relative to some frame, and Q_2 is simultaneous with P_1 relative to some other frame. This tells us that tenselessly, part P_2 of O is just as ontologically real as part P_1, but the three-dimensionalist never denied that.

So I think it is far more likely that the three-dimensionalist will accept (5), but will deny that simultaneity is merely sufficient for co-existence and thus will deny (4), that co-existence is transitive. Since the three-dimensionalist notion of being wholly present is defined in terms of co-existence, the definition of co-existence that is used must be one that captures what the three-dimensionalist means by the term. But it seems clear that when the three-dimensionalist says that an object is wholly present at t just if all of its parts (S-*simpliciter*) co-exist at t, she means that an object is wholly present at t just if all of its parts (S-*simpliciter*) exist simultaneously at t.

To illustrate, let us consider an example. Suppose there is a man M who presses a red button and at that moment is vaporised. Pressing the button then causes a machine, to a moment later, instantly create a woman from the surrounding atomic matter. Suppose I am observing these events from some inertial reference frame. This is represented in the diagram below, where M represents a point on the worldline of the man, and W represents a point on the worldline of the woman. The point W at which the woman comes into existence is in the absolute future of the point M at which the man ceases to exist. The diagonal lines represent everything that is outside the future and past light cones of me relative to my reference frame.

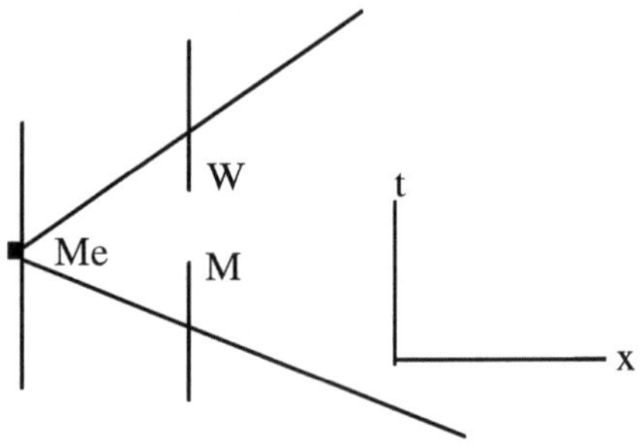

Relative to my reference frame, I am simultaneous with both the man when he is still alive prior to pressing the button, and with the woman after she is miraculously created. Thus I co-exist with both the man and the woman. If co-existence is transitive, then it follows that the man co-exists with the woman, despite the fact that the man ceases to exist before the woman comes into existence: the man's ceasing to exist is in part the cause of the woman coming into existence. There is no sense in which the man and the woman exist at the same time. This is because simultaneity is not transitive. Though relative to my frame of reference I am simultaneous with both the man and the woman, it does not follow that the man is ever simultaneous with the woman. So it is difficult to see why we should think there is any sense in which the man and the woman co-exist.

This is not to say that the (eternalist) three-dimensionalist cannot embrace the minimalist 'block universe' sense of co-existence according to which it is tenselessly true that everything co-exists with everything else. She could call this sense of co-existence, capital '*C*' Co-existence. But she will not define being wholly present at a time in terms of *this* notion of Co-existence. Rather, the sense in which the parts of persisting objects co-exist, is a sense of co-existence according to which simultaneity is necessary for co-existence, and where co-existence is therefore intransitive.

That co-existence is intransitive is not a peculiar feature of a three-dimensionalist account of co-existence. Suppose we are perdurantists. Then we might say that there is *some* notion of co-existence that is captured by the minimalist block universe conception of co-existence. Along with the three-dimensionalist, we might call this notion 'Co-existence'. Presumably though, the perdurantist will also want to capture some non-trivial sense of co-existence according to which I co-exist with President Bush, but do not co-exist with Atilla the Hun.

The perdurantist might say that any two perduring objects co-exist just if each has temporal parts that exist at the same time relative to some reference frame.[10] Specifically, he might say that:

> *O* and *O** co-exist just if there exists some synchronic fusion *F* that is a temporal part of *O*, and some synchronic fusion *F** that is a temporal part of *O**, and relative to some frame of reference *R*, *F* and *F** exist simultaneously.

In this sense of co-existence, it is not even true that co-existence is transitive *relative to one and the same frame of reference*. Suppose that relative to his rest frame, R_1, my grandfather exists from t_1 to t_{10}, and relative to my rest frame, also R_1, I exist from t_7 to t_{20}. Suppose also that relative to R_1, my dog exists from t_{11} to t_{18}. Then there is some frame of reference, R_1, from the perspective of which one of my temporal parts is simultaneous with one of my dog's temporal parts, (at t_{11}) and from the perspective of which one of my temporal parts is simultaneous with one of my grandfather's temporal parts (at t_8). So I co-exist with my dog, and I co-exist with my grandfather. But even though each of these simultaneity relations occurs relative to the same frame of reference, (R_1) it does not follow that my dog co-exists with my grandfather. For it does not follow that there exists any temporal part of my dog that is simultaneous with any temporal part of my grandfather, and indeed, there is not, since my grandfather dies before my dog is born.

This sense of co-existence is what we might think of as a non-trivial tenseless sense. It is the sense in which if it is ever true that I co-exist with some object *O*, then it is tenselessly true that I co-exist with *O*. It tenselessly true of me that I co-exist with my grandfather, and thus is true of me even at times when my grandfather is no longer alive.

But there is clearly another sense of co-existence. Rather than talking of four-dimensional wholes tenselessly co-existing in virtue of having some simultaneously existing temporal parts, perdurantists could talk of four-dimensional wholes co-existing at times, in virtue of having simultaneously existing temporal parts at those times. If we suppose that the current time is t_{13}, then while it is true that *some* temporal part of me co-exists with *some* part of my grandfather, it is not true that my *current* temporal part—*I-now*—co-exists with any part of my grandfather. *I-now* is in the absolute future of all temporal parts of my grandfather. So, we might say, relative to some reference frame *R*, a perduring object *O* co-exists with *O** at *t* just if *O* and *O** both have temporal parts that exist at *t*. That is,

> *O* co-exists with *O** relative to *R* at *t*, just if relative to *R*, at *t* there exist synchronic fusions *F* and *F** such that *F* is a temporal part of *O* at *t* and *F** is a temporal part of *O** at *t*.

(Relative to any frame of reference *R*, the *t*-parts of any two persisting objects will exist simultaneously: that is what it is to be the *t*-part of an object. But relative to different frames of reference, different temporal parts will count as *t*-parts for any given *t*.) This account of co-existence mentions both frames and times—it indexes co-existence to times and frames. Then it is tenselessly (and framelessly) true of *O* and *O**, that they co-exist at the times relative to the frames, that they do. That is, it is tenselessly true of me, that relative to R_1 I co-exist with my grandfather at t_7: that is true of me even at t_{18}. But again, co-existence is not transitive. Although there may be some frame of reference *R** from the perspective of which *I-now* is simultaneous with *O*, and from which *O* is simultaneous with some temporal part of my grandfather, it does not follow that *I-now* co-exists with any temporal part of my grandfather: *I-now* is never simultaneous with any temporal parts of my grandfather.

The distinction between these two non-trivial senses of co-existence bears a resemblance to the earlier distinction we drew between the M- and S-*simpliciter* senses of *simpliciter*. It is tempting to say that the former tenseless sense of co-existence is the M-*simpliciter* sense of co-existence such that:

> *O* and *O** co-exist M-*simpliciter* just if there exists some synchronic fusion *F* that is a temporal part of *O*, and some synchronic fusion *F** that is a temporal part of *O**, and relative to some frame of reference *R*, *F* and *F** exist simultaneously.

We might think that the latter indexed sense of co-existence is not quite the S-*simpliciter* sense of co-existence. For the S-*simpliciter* notion is a tensed notion. The utterance '*O* is red S-*simpliciter*' may be true at one time, and false at another—it will be true when the manner in which red is instantiated, is the same as the current time and false otherwise. But of course, if *O* is red S-*simpliciter* at t_1, then it is tenselessly true that *O* is red S-*simpliciter* at t_1. Thus we can see that our indexed notion of co-existence really is the S-*simpliciter* sense of co-existence, according to which *O* and *O** co-exist S-*simpliciter* at a frame and time, just if *O* and *O** have temporal parts that are simultaneous relative to that frame at that time. Hence my grandfather and I co-exist S-*simpliciter* at R_1 at t_7. This is tenselessly true. But the utterance 'my grandfather and I co-exist S-*simpliciter*' made relative to R_1 at t_7 is true, and made relative to R_1 at t_{11} is false. Hence we will say that:

> *O* co-exists with *O** S-*simpliciter* relative to *R* at *t*, just if relative to *R*, at *t* there exist synchronic fusions *F* and *F** such that *F* is a temporal part of *O* at *t* and *F** is a temporal part of *O** at *t*.

Notice that both of these non-trivial intransitive senses of co-existence are ones that the three-dimensionalist can embrace, albeit in slightly modified form. Specifically, the endurantist will say thc following:

> *O* and *O** co-exist M-*simpliciter* just if relative to some frame of reference *R*, there exists some synchronic fusion *F* that constitutes *O* at some time, and some synchronic fusion *F** that constitutes *O** at some time, and relative to *R*, *F* and *F** exist simultaneously.

And:

> O co-exists with O^* S-*simpliciter* relative to frame R at t, just if relative to R, at t there exist simultaneous synchronic fusions F and F^* such that F constitutes O at t, and F^* constitutes O^* at t.

So in general, persisting non-unitary objects O and O^* co-exist M-*simpliciter* iff relative to some reference frame R, there exists some time at which synchronic fusion F exists and is related to O, and there exists some time at which synchronic fusion F^* exists and is related to O^*, and F and F^* exist simultaneously relative to R. Persisting non-unitary objects O and O^* co-exist S-*simpliciter* relative to R at t, just if relative to R, at t there exist synchronic fusions F and F^* such that F is related to O at t, and F^* is related to O^* at t.

In the case of unitary versions of three- and four-dimensionalism, no appeal can be made to the existence of synchronic fusions that are simultaneous relative to certain frames. So instead, unitary views will have to embrace the locution of persisting objects existing at a time relative to a reference frame. (Though for three-dimensionalists, persisting objects exist at a time relative to a frame in virtue of being wholly present at that frame and time, while for four-dimensionalists they exist at a time and frame in virtue of some 'hunk' of the object existing at that time and frame). Hence both can adopt the same account:

> O and O^* co-exist M-*simpliciter* just if relative to some frame of reference R, there exists some time t at which O exists, and some time t^* at which O^* exists, and t is identical to t^*.

and:

> O and O^* co-exist S-*simpliciter* relative to R at t, just if relative to R, O exists at t and O^* exists at t.

Thus there are analogous three- and four-dimensionalist accounts of co-existence M- and S-*simpliciter*. On none of these non-trivial accounts, however, is co-existence transitive. Yet Hales' and Johnson's argument requires that we move from the claim that P_1 co-exists with Q_2, and Q_2 co-exists with P_2, to the conclusion that P_1 co-exists with P_2, and that follows only if co-existence is transitive. Thus the argument fails. But although their argument may not show that special relativity is strictly speaking inconsistent with three-dimensionalism, it may nevertheless be the case that special relativity and three-dimensionalism are uncomfortable bed fellows. This is the contention of the moderate view.

2. THE MODERATE VIEW

2.1 Co-existence and Endurance

Suppose we have two persisting objects *O* and *O**. Consider *O*'s worldline, and consider some point, *O1*, on that worldline. Now consider *O**'s worldline, and consider two points on that worldline, *O*1* and *O*2*.

Suppose that relative to *O's* frame of reference at *O1*, *O* is simultaneous with *O*1*, and with *O*2*. Then *O* co-exists with *O** more than once: it co-exists with *O** when *O** is at *O*1*, and when it is at *O*2*. According to Yuri Balashov, if *O* and *O** are enduring objects, then this is counterintuitive in a way that it is not counterintuitive if they are perduring objects.[11] For if *O* and *O** perdure, then *O*1* and *O*2* are two distinct temporal parts of *O**. So when *O* co-exists with both *O*1* and *O*2*, it is merely co-existing with two distinct objects that happen to each be part of one and the same perduring object. Similarly, since terduring objects are spread out in time, there is nothing mysterious about *O* co-existing multiple times with *O** as *O** exists at different spatio-temporal regions. But if *O* and *O** are enduring objects, then *O** is wholly present at *O*1* and *O*2*, and is strictly identical to itself at each of those times. But how can *O* co-exist with *O** twice at the same time if *O** is wholly present whenever it exists; how can *O*1* and *O*2* be one and the same object if they are 'both' co-existent with *O* at *O1*?

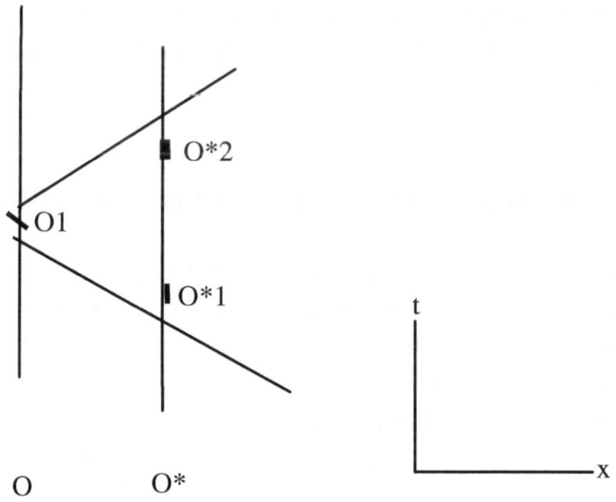

There is no denying that our pre-relativistic intuitions tell us that *O** co-existing with *O* at *O1* twice is odd. But our intuitions tell us that even if *O* and *O** are perduring or terduring objects. *O*2* is in the absolute future of

*O*1*. If we suppose that *O** perdures, then it is not merely that *O*1* and *O*2* are distinct objects that just happen to be temporal parts of some perduring *O**. Rather, (assuming *O** is a normal persisting object) *O*2* exists only because *O*1* exists: *O*2* exists as the result of *O**'s causal propagation across time. *O*1* immanently causes *O*2* to exist. But if *O*1* causes *O*2*, then it is certainly counterintuitive to think that *O* can co-exist with both *O*1* and *O*2*.

Putting aside the fact that multiple co-existence of this kind is counterintuitive, however, I see no reason to suppose that this scenario is any *more* counterintuitive for the three-dimensionalist than for the four-dimensionalist. After all, every three-dimensionalist holds that persisting objects wholly exist at multiple spatio-temporal regions: they are wholly present at every point on their worldline. As we saw in the previous chapter, this means that according to the three-dimensionalist, if an object travels back in time to meet its younger self, then a single enduring object is wholly present at two spatial regions at the same time. And that object is strictly identical to itself at each of those regions. Yet this is not counterintuitive (it is presumably how most folk conceptualise time travel scenarios). So we know that an enduring object can wholly exist at two spatio-temporal regions that are, from the perspective of their rest frames, simultaneous. So even if it had turned out that *O*1* and *O*2* exist simultaneously, this would present no problem for the three-dimensionalist. But of course, they do not. *O*1* and *O*2 themselves* never co-exist: one is in the absolute future of the other. It is simply that *O* at *O1* co-exists with *O** twice. There is simply no reason, though, why this should be any problem for the three-dimensionalist.

2.2 Special Relativity, Parthood and Properties

Is there some other problem that special relativity raises for the three-dimensionalist? According to Hales and Johnson, the three-dimensionalist is faced with a dilemma. If she holds that an enduring object *O* is composed of *all* of the parts that compose *O* relative to each possible frame of reference, then she is committed to holding that *O* is composed of parts that, from the perspective of one or more frames of reference, are not simultaneous. Then she must reject the idea that *O* is wholly present. On the other hand, if she holds that *O* is composed of the parts that are simultaneous in a particular reference frame at a particular time,[12] then, they argue, she is lead into unacceptable metaphysical profligacy. For in that case, the object composed of the parts that are simultaneous relative to a reference frame *R*, is not the same object as the object composed of the parts that are simultaneous relative to reference frame R^*. For each of *these* objects is composed of different parts at different times, and must therefore be distinct.

Three-dimensionalists will no doubt reject the former view in favour of the latter: enduring objects are composed of parts that are simultaneous relative to a particular reference frame at a particular time. It is this that has lead Balashov to describe three-dimensional objects as 'relativistic',[13] and has led defenders of perdurantism, such as Smart[14] and Quine,[15] to reject three-dimensionalism on the grounds that the same enduring object will have different temporal and spatial properties depending on the frame of reference of the observer, and this, they think, is inconsistent with such objects being wholly present whenever they exist.

Consider the following example modified slightly from Hales and Johnson.[16] Suppose Dave is sitting in the middle of a very high speed train which travels at a significant fraction of the speed of light. At each end of the train there is a clock, and Dave has synchronised his watch with each of the clocks on the train. Suppose the train has just two carriages, a front and rear carriage. At each end of the train there is a switch, which if pressed, results in that half of the train almost immediately changing colour to red inside and out. At each end of the train there is a man sitting next to the switch. Dave is in the rest frame of reference of the train, and from his perspective at 12.00 exactly, both switches are pressed, and just after 12.00 the entire train is red. From Dave's perspective, prior to 12.00 the train was a uniform silver, and post 12.00 the train is a uniform red. From Dave's perspective, there is no time at which one half of the train is red and the other half silver.

Suppose that as the train passes a station, Sally is standing on the platform. Suppose Sally is directly opposite Dave when his watch reads 12.00. At that time, from Sally's frame of reference, the front half of the train appears a uniform silver, while the back half of the train is red. According to Sally, the train is at one time half silver and half red. Thus Sally is committed to the existence of a train that is at one time half red and half silver, while Dave is committed to the existence of a train that is at all times a uniform colour. Balashov, Smart and Quine conclude that given three-dimensionalism, persisting objects are unacceptably relativistic—relative to one frame the train is a uniform colour, and relative to another frame it is not a uniform colour, despite the fact that the train is supposed to be wholly present whenever it exists. Hales and Johnson, on the other hand, conclude that three-dimensionalism leads to unacceptable metaphysical profligacy because there are really *two* trains, one that is always a uniform colour and one that is at one time half red and half silver.

The three-dimensionalist can accept that enduring objects will be composed of different parts and have different properties relative to different frames of reference, but should point out that this does not lead to metaphysical profligacy, nor is it impossibly counterintuitive. For there is nothing contradictory about having different properties or parts relative to different frames of reference. For it is not that there is some absolute time t at which, from one reference frame an object O has a part P, and from another frame lacks P. So it is not that there is some absolute time t, at which there is a train that is half red and half silver (from one reference frame) and a train that is wholly red (from another reference frame). That might suggest that we have two distinct trains. But there is no absolute time independent of frame of reference. Rather, an object has some part P at a *different time relative to a different inertial frame*, not at a different time *simpliciter*.

We can see this problem as a relativistic analog of the problem of temporal indexing from within a frame of reference. How is it that from within a frame of reference R, one and the same enduring object can have different properties at different times and yet be the same object at each of those times? The answer is that the object has those properties in a temporally adverbialised manner: $t$ly. But it is still true that one and the same object has the property of being, say, red $t$ly, and that it has this property *at times other than t*. So too we can say that an enduring object has properties 'frame-of-referencely' or Rly. An enduring object has some property P at t Rly if it has that property at t relative to R. So the train has the property (in some t_1 ly manner) of being a uniform silver Rly where frame R is the rest frame of the train (Dave's frame), and it has the property $t_2$ly, of being a uniform red Rly. It also has the property $t_2$ly of being half red and half silver R*ly, where R* is the frame from which Sally observes the train.

So it is true of the train at every time and at every frame that it has all of these properties. It is true at the time at which Sally observes the train, that it has not only the property of being half red and half silver R*ly, but also that it has the property of being a uniform silver $t_1$ly Rly, and of being a uniform red $t_2$ly Rly. This is why the train is strictly identical across times and frames. Thus there is no reason to posit the existence of multiple trains that are composed of different parts relative to different frames: there is one train that has all of the same frame invariant 'framely' properties at all frames and times, in the same way that relative to a frame, there is a single object that tenselessly has all of the same temporally adverbialised properties.

The sense in which three-dimensional objects have different properties relative to different frames is the same as the sense in which relative to a single frame, they have different properties relative to different times. It is certainly true that enduring objects *manifest* different properties at different times—this is the sense in which they change over time—but since they are strictly identical across time, it is also true that they have the *same* properties at each time at which they exist. It is always true of some object O that is red at t_1, that it is red $t_1$ly, even at t_2 when it is manifestly green. And just as being red $t_1$ly at t_1 is being manifestly red, so too being P frame Rly at frame R, is being manifestly P. So although three-dimensional objects manifest different properties relative to different frames, they nevertheless instantiate the same framely properties at every frame.

But how does this three-dimensionalist account compare with a four-dimensionalist account? In effect, for the four-dimensionalist what is happening on these occasions is that we are seeing one and the same four-dimensional object from different 'angles' in space-time. For the perdurantist, if we think of four-dimensional objects as being like sausages, then depending on how we slice the sausage we get different shaped sausage slices with different properties. Usually we slice the sausage 'head on', so that each slice is round. These three-dimensional slices represent the maximal temporal parts of a four-dimensional object from its rest frame. If we slice the sausage on a slight angle we get elongated slices that have different properties to the round slices. These elongated slices represent the maximal temporal parts of a four-dimensional object from some inertial frame of reference. So for the perdurantist, one and the same four-dimensional object will have different maximal temporal parts relative to different frames of reference: for temporal parts just are those objects that overlap all of the simultaneous spatial parts of a four-dimensional object at a time. Since different parts are simultaneous relative to different frames, it follows that perduring objects will have different temporal parts relative to different frames. But these temporal parts are simply different three-dimensional slices of a single four-dimensional object. (Though frequently these slices will be odd gerrymandered objects that are of little interest to us).

In the case of the train, there are (maximal) temporal parts relative to one frame of reference, R, such that one temporal part is wholly silver, and the temporally contiguous part is wholly red. Relative to R^* though, there is a temporal part that is half red and half silver. That is because these parts represent different slices of the four-dimensional train.

Are these accounts substantially different? I say not. According to the perdurantist, what you see from different frames of reference are different temporal parts of one and the same object. We have already noted that if some four-dimensional object O is red at t in virtue of having some temporal part at t that is red *simpliciter*, then in order to talk about the properties of O as a whole, we can talk about the properties the whole instantiates $t$ly. The same will be true when we consider temporal parts across different frames of reference. One temporal part of the train is half red and half silver. Other temporal parts of the train are a uniform colour. The whole train, however, cannot be both uniformly coloured and dual coloured. So to talk about the properties of the whole train, we need to talk of its properties relative to a frame of reference, or *R*ly. So the perdurantist can say that the whole train is dual coloured *R**ly, and is uniformly coloured *R*ly, in virtue of having some temporal part relative to *R* that is uniformly coloured, and some temporal part relative to *R**, that is dual coloured.

Ultimately, both three- and four-dimensionalists will think about persisting objects relative to different frames of reference, in an analogous way to the way in which they think about persisting objects relative to times. For endurantists and perdurantists, talk from within a frame of reference, of synchronic fusions constituting or being temporal parts of persisting objects at times, will be expanded to include talk of synchronic fusions constituting or being temporal parts of persisting objects relative to frames of reference. Both endurantists and perdurantists agree that relative to a particular frame of reference, there exist at different times, distinct three-dimensional objects (synchronic fusions) that are related to persisting objects at times. They also agree that relative to different frames and times, there exist distinct synchronic fusions. For the perdurantist, these fusions represent different temporal parts relative to different frames, of a single four-dimensional object. For the endurantist these fusions constitute, relative to different frames, a single enduring object. Hence any (non-unitary) persisting object *O* instantiates properties *t*ly *R*ly, just if relative to *R*, at *t* there exists some synchronic fusion *F* that is related in some manner to *O*: *F* is a temporal part of *O* at *t*, or *F* constitutes *O* at *t*.

For unitary three- and four-dimensionalists, the various three-dimensional 'slices' that we see from different frames of reference are not synchronic fusions that are related in some way to persisting objects. Rather, according to the unitary three-dimensionalist there exists a single enduring object that is wholly present at each time relative to each frame of reference at which it exists. This enduring object has a single set of frame invariant *R*ly properties and tenseless *t*ly properties that it instantiates at all times relative to all frames. Hence it is strictly identical across frames and times. Similarly, for the unitary four-dimensionalist there is a single terduring object which, viewed from different frames of reference, has different properties. But the properties of the whole terduring object are frame invariant. Any terduring object has the property, from every frame of reference, of having certain properties relative to certain frames of reference: if terduring object *O* has *P* at *t* relative to *R*, then *O* has *P* *t*ly *R*ly at every time relative to every frame.

Given this account of the properties of persisting objects across different frames, we can see that the three-dimensionalist is able to cope with special relativity without incurring costs of metaphysical profligacy, and indeed, without objects turning out to be 'relativistic' in any strong or controversial sense that renders three-dimensionalism more counterintuitive than four-dimensionalism. Moreover, we call see how the translation functions that we developed in chapters five and six, will be expanded in the light of considerations pertaining to special relativity and the nature of space-time more generally. For each three-dimensionalist theory treats across-frame property differences in the same manner they treat across-time property differences, as does each four-dimensionalist theory. Then if we can successfully translate across-time property talk of analogous three- and four-dimensionalist theories, we can apply the same strategy for across-frame property talk.

3. THE WEAK VIEW

Finally, we can consider the weak view. The core argument for this view is due to Yuri Balashov, who argues that given the four-dimensional geometry of Minkowski space-time, four-dimensionalism has explanatory resources that three-dimensionalism lacks.[17] Both there- and four-dimensionalists agree that persisting objects occupy a particular volume in spacetime. Furthermore, both agree that there exist various three-dimensional objects, and that those objects have different properties relative to difference frames of reference. The perdurantist, however, holds that the four-dimensional volume is occupied by a four-dimensional object that is the mereological fusion of the various three-dimensional objects which are instantaneous temporal parts of that persisting object. The three-dimensionalist holds that each of the three-dimensional slices of the four-dimensional volume is occupied by a wholly present object which is strictly identical across the four-dimensional volume.

For the perdurantist then, there exists a relativistically invariant object—the four-dimensional whole—and observing this invariant object from different perspectives generates the various three-dimensional shapes. Think once again of the four-dimensional volume as a sausage. Slice the sausage along different planes and you get different shaped sausage slices. What explains why those slices have the properties they do is that they were sliced from the particular sausage they were, in the particular manner they were. So while the three-dimensional objects exemplify different properties relative to different frames of reference, there is some objective, invariant shape that 'stands behind', and thus explains each of these different shapes. But what stands behind the relativistic three-dimensional objects if one is a three-dimensionalist? As Balashov puts it, the three-dimensionalist has to start with the three-dimensional objects, and then discover that they can be arranged into a unified four-dimensional volume, and this arrangement must be nothing more than brute fact.[18]

This way of thinking of things is a bit misleading. It brings to mind the image of finding variously shaped bits of meat around one's house, and then discovering that they fit together into a sausage shape, despite the fact that they are not 'sausage-slices'. Consider this idea of a 'nice four-dimensional volume' more closely. For perdurantists who believe in unrestricted mereological composition, not all four-dimensional objects will fill nice four dimensional volumes—in fact, most won't. Why do the fusions of some temporal parts fill nice volumes? Because they are the temporal parts of a causally self-propagating object, an object whose temporal parts are related by immanent causation, and as such are spatially and temporally contiguous. Why do these three-dimensional objects fit together to form a nice volume if one is a three-dimensionalist? Well, when they do nicely fit together they fit together for exactly the same reason: they are not distinct three-dimensional objects that just happen to fit together, rather, the four-dimensional volume just is the entire lifespan of the enduring object that fills that volume, and it neatly fills that volume because it is causally related to itself at every time at which it exists. Some enduring objects propagate themselves through time, and by doing so they fill nice four-dimensional volumes. There is no explanatory mystery here.

Indeed notice something here. If all we had were relativistic three-dimensional shapes and no theory about how they 'fit together', we might be surprised to discover that they fill the volumes they do. The theory of special relativity, however, along with various other laws of nature, allows us to predict how objects that exist in the present, will exist in the future. That is, they allow us to predict what the four-dimensional volume of an object will be. For in fact, we do not take a bird's eye view of the universe and see four-dimensional objects, which we can then use to explain the various frame relative shapes. Rather, we take as basic the three-dimensional objects, and use the various 'rules' in the form of the laws of nature, to predict what those objects will be like in the future. So it can hardly come as a surprise when we discover that those objects fill nice four-dimensional volumes: for that is precisely what we predicted.[19]

So both three- and four-dimensionalists will predict that some persisting objects will fill 'nice' four-dimensional volumes, and both have the resources to explain why this is so. There is, therefore, no explanatory virtue of four-dimensionalism over three-dimensionalism. Indeed, so far we have found neither the strong, the moderate nor the weak views to be plausible. Instead, we have seen that analogous versions of three- and four-dimensionalism are equally consistent (or perhaps inconsistent in the case of presentist versions) with special relativity, and that each have the same explanatory resources. So far then, the empirical equivalence of three- and four-dimensionalism remains unshaken, and the case for their metaphysical equivalence stronger still. There is, however, one last matter to address: the issue of the relation between objects, between space-time regions, and between objects and space-time regions.

4. SUBSTANTIVALISM/RELATIONALISM; MONISM/DUALISM

Throughout, I have spoken of persisting objects *occupying* space-time regions. But this glosses over a number of distinctions one might make between the relations between objects and the relations between space-time regions on the one hand, and the relations between objects and space-time regions on the other hand. Substantivilists are those who are realists about space-time regions and, if they exist, space-time points. Relationists hold that space-time regions and points are not ontologically real. Talk of space-time regions is not talk about some entity, a space-time region, but rather is really just talk about locations in space-time, which is ultimately reducible to talk about the relative spatial and temporal distances between objects.

There is another distinction that cuts across the substantivilist/relationalist distinction: the monist/dualist distinction.[20] Monists hold that the relation between space-time regions and the objects that exist at those regions[21] is one of identity. Dualists hold that there exist ontologically real space-time regions such that some of those regions are occupied by objects which are distinct from those regions. Dualists are necessarily substantivilists. Monism, on the other hand, comes in two flavours: substantivilist and relational. Substantival monists think that space-time regions are ontologically real, and that objects are to be identified with those regions or the pattern of properties possessed by them. Relational monists think that only objects exist, and to the extent that we can talk about space-time regions, we are really just talking about those objects and the relations between them.

What has this to do with the equivalence of three- and four-dimensionalism? Notice that the four-dimensionalist can coherently embrace any of these three views: substantival monism, substantival dualism and relational monism. Three-dimensionalists cannot. For any enduring object *O* exists at multiple space-time regions $R_1 \ldots R_n$ and is identical to itself at each of those regions. But if space-time regions are ontologically real, then no two space-time regions are strictly identical, and hence $R_1 \ldots R_n$ are distinct. Thus by transitivity of identity, *O* cannot be identical to any of the distinct regions at which it exists. Hence three-dimensionalism is inconsistent with substantival monism.

Three-dimensionalism is consistent with both substantival dualism, and with relational monism. (Even in this latter case it makes sense to talk of objects occupying space-time regions, so long as this is properly understood in a relationalist manner. Enduring objects are strictly identical to themselves across time. Yet we want to say that they wholly 'occupy' different regions of space at different times. This sounds a little odd, since it entails that the claim that region *R* is identical to region *R** just if *R* and *R** wholly contain all and only the same object, is false. For each region at which an enduring object exists, wholly contains all and only that object, yet each of those regions is distinct, and is so in virtue of being differently related to various other objects.)

Does this give us reason to suppose that three- and four-dimensionalism are not equivalent? Perhaps. One could make a stand for non-equivalence based on these considerations. In response, one could argue that substantival monism is not just false but necessarily so: the relation between objects and the regions at which they exist is contingent, as is attested by the fact that each have different modal properties. Objects could have existed at regions other than the regions they in fact exist at. But this objection is not decisive. It is true that those substantivilists who think that Lumpl and Goliath are distinct and necessarily so, should for analogous reasons reject monism in favour of dualism. As Sider points out, however, just as an appeal to counterpart theory and contingent identity can resolve the Lumpl and Goliath puzzle, so too can it resolve this problem.[22] Though the very same things are picked out by the designation 'object' as by 'space-time region', nevertheless each of these designations have attached to them different sets of counterpart relation, region-relations and object-relations, such that not all of the region-counterparts are also object-counterparts and *vice versa*, thus explaining why objects and regions have different modal properties.

Where does this leave the three-dimensionalist? Consider non-unitary theories. According to the perdurantist who embraces substantival monism, there exist four-dimensional regions of space-time that are identical to the four-dimensional objects that exist at or 'within' that region. Call one such region *R*, and one such object *O*. Then each synchronic fusion that is a maximal proper temporal part of *O*, is identical to some three-dimensional region of space-time that is a proper sub-region of *R*. Hence just as the relation between objects and their temporal parts is the part/whole relation so too is the relation between regions and their sub-regions. In general then, for each sub-region $R_1 \ldots R_n$ of *R*, there is some proper part of *O* that is identical to that region.

Notice that the endurantist can agree that there exist synchronic fusions that are identical to the three-dimensional regions of space-time at which they exist. What she denies is that enduring objects are *themselves* identical to any of these synchronic fusions, and hence identical to these regions. Rather, enduring objects exist at different spatio-temporal regions, in virtue of being constituted at those regions, by synchronic fusions that are identical to those regions. Now notice something further. If endurantism and perdurantism are equivalent, as I have argued, it is because talk of being constituted at times by synchronic fusions, is equivalent to talk of having those fusions as parts at times. Ultimately, this means that the three-dimensionalist sense of enduring objects being strictly identical to themselves at every time, is equivalent to the four-dimensionalist sense of perduring objects failing to be strictly identical to themselves at every time: identity according to one view, is non-identity according to the other view. But if that is true, then it is no further leap to think that being constituted by a fusion *F* at a time *t* that is identical to some region R_1, and having as a part some fusion *F* at *t* that is identical to some region R_1, are also equivalent. Perdurantist talk of being identical to a four-dimensional region, is equivalent to endurantist talk of occupying a four-dimensional region. Does that mean that those who are committed to the equivalence of non-unitary three-and four-dimensionalism, must also be committed to the equivalence of substantival monism and dualism? No. For substantival dualism holds that it is always the case that space-time regions and the objects that exist at those regions, are distinct. Substantival monist endurantists, however, hold that synchronic fusions are identical to the regions at which they exist. The only context in which talk of occupancy and identity is equivalent, is when talking about the persisting objects that are related in some manner to those fusions.

What of unitary versions? Here matters are more difficult. For there are no synchronic fusions that coincide with persisting objects at times, which are candidates for being identical to the space-time regions at which they exist. Since all three-dimensionalists must deny that enduring objects are identical to any of the three-dimensional regions at which they exist at a time, it seems that the unitary three-dimensionalist must reject substantival monism outright. *Prima facie* though, it looks as though the unitary four-dimensionalist can accept substantival monism, insofar as she can coherently hold that an entire terduring object is identical to the four-dimensional space-time region at which it exists. And if that is so, then the claim that unitary three- and four-dimensionalism are equivalent looks to be on shaky ground.

But there is a problem. Consider some terduring object *O*. Suppose *O* fills four-dimensional region *R*. *O* has various parts: if it is a terduring non-fusion, then it has various spatial parts at times (the parts of the fusion(s) that compile it at those times); if it is a terduring fusion, then it tenselessly has as parts the terduring simples that it fuses. Given substantival monism, each of these parts of *O* is identical to some sub-region of *R*. All well and good. Given standard views about space-time regions however, *R* also has various other sub-regions that are proper parts of *R*: for instance, there are various three-dimensional regions that are sub-regions of *R*. Some of these three-dimensional sub-regions are identical to spatial parts of *O*. But some are not. According to the *perdurantist*, the three-dimensional regions that are not identical to spatial parts of *O*, are identical to (maximal or non-maximal) temporal parts of *O*. But the unitary four-dimensionalist denies that *O* has any temporal parts. Yet if *O* is identical to *R*, then *R* cannot have proper parts that *O* does not. So it looks as though the unitary four-dimensionalist must either say that certain four-dimensional regions are such that there 'are' certain sub-regions of those regions that do not exist, or that those sub-regions exist, but are not parts of the regions in question.

Now, there is clearly a sense in which the latter option cannot be quite right. For it is not simply that unitary four-dimensionalists deny, as do endurantists, that there exist instantaneous objects that overlap persisting objects at times, and are parts of those objects at those times. That is, unitary (three- and) four-dimensionalists do not simply deny that certain three-dimensional sub-regions of *R*, are *parts* of *O* (and *R*). Rather, she denies that there exist at those sub-regions, any distinct object that is related *in any way at all*, to the persisting object *O* that exists at *R*. That is, she denies the following:

> For every persisting object *O*, if *R* is the volume of space-time at which *O* exists, and sub-*R* is any sub-region of *R* at which some object can exist, then there exists some object that exists at and only at sub-*R*.

But this denial is defined in terms of the existence of sub-regions. And given substantival monism, doesn't it follow that some of these sub-regions of *R* fail to exist? That depends what you think it takes to count as a sub-region. There is a sense in which the unitary four-dimensionalist can talk of sub-regions. Presumably what it is to talk of some sub-region of *R*, is really to talk of *R at particular space-time points*. When we talk of some terduring object *O* at a time, we are not talking about some temporal part of the object that exists at that time—that is, some three-dimensional object that exists at and only at a particular three-dimensional sub-region of *R*. We are talking about *O* at a time and location: *O* at a particular set of space-time points. But there is no distinct object that exists at and only at those points and that is a part of *O*.

None of this is at all heartening though. It is difficult to know what it means to talk about a space-time region at a point. It is difficult to know what to say about the difference between sub-regions of regions that are proper parts of that region, and pseudo sub-regions of regions that are not parts of the region, but are the region 'at a particular set of points'. It is difficult to know what to say about the relation *between* each of these pseudo sub-regions, given that each appears to be distinct and to have different properties, and yet each is *the very same region* at a different set of points.

Indeed, those who hold that there exist spatially and/or temporally extended partless objects naturally also hold that those objects are not identical to the space-time regions at which they exist. We have seen why. Given these considerations, I think we can say that the unitary four-dimensionalist is no better able to embrace substantival monism than is the unitary three-dimensionalist.

What we have seen is that both non-unitary three- and four-dimensionalism can coherently accept all three views about the relation between space-time regions and objects. We can also see that unitary three- and four-dimensionalists alike will be unable coherently to accept substantival monism. So consideration of these views merely reinforces the claim that analogous versions of three- and four-dimensionalism have the very same resources and are consistent with the very same ancillary commitments.

5. CONCLUSION

It turns out that our world is not the nice Newtonian absolutist world that we have always thought, and which our intuition resolutely insists on telling us it is. When we find out the way the world actually is, we find that it is a way that does not preserve all of our pre-theoretic intuitions. But this is true whatever one's theory of persistence.In fact, what we see is that analogous versions of each theory cope in the same manner with the empirical reality of our universe, though neither preserves all of the pre-relativistic intuitions with which we began. So any two theories of persistence, or metaphysical packages, that differ only with respect to the component of persistence—that is, which differ only with respect to whether they hold that objects are three- or four-dimensional—are empirically equivalent. What's more, we have good reason to suppose that they are correctly inter-translatable. Whatever puzzles we throw at them, each is equally able to resolve those puzzles in an analogous manner. These theories are thus equally explanatory in a strong sense: they not only explain all the same phenomena, but the explanations they provide are of the same kind. Hermeneutic considerations tell us that in general we should think that proponents of each theory speak the truth: that is, proponents of one theory do not constantly speak falsehoods. Ultimately, inference to the best explanation tells us we should conclude that analogous versions of three- and four-dimensionalism are equivalent.

Previously we might have thought that what distinguishes one metaphysical package—one complete theory of persistence—from another, is a difference regarding the dimension of persistence—that is, a difference regarding whether strictly speaking objects are three-, or four-dimensional. We now know, however, that this is not so. There can be no difference in the dimension of persistence in *this* sense. For metaphysical packages that do not differ with respect to any components except the dimension of persistence, do not differ at all. So the issue of how we should think about objects at and across time is not the issue of whether we should think that persisting objects are three-dimensional or four-dimensional. Nor is it the issue of whether we should think that such objects are strictly identical across time, nor whether we should think that they are composed of temporal parts. For these are not genuine issues: they are mere artefacts of the different languages spoken by the three- and four-dimensionalist. But we cannot see that these are different languages until we avail ourselves of the resources of an account of inter-theoretic equivalence. Only then do we realise that there are cases where what looked like an issue of genuine metaphysical contention, turns out to be a matter of simple semantic disagreement.

We might say then, that this book embodies a particular methodology: the methodology of bringing second-order metaphysics to bear on first-order problems. Indeed, its key insight is that first- and second-order metaphysics are deeply inter-related, and that any attempt to answer first-order questions without first settling second-order issues, will frequently lead us astray. One of the fundamental conclusions of the book is that we have been lead astray in just this manner in the domain of the first-order metaphysics of persistence. For applying this methodology to the metaphysics of persistence yields substantial conclusions that challenge a number of widely held views about objects and their persistence. These conclusions challenge the claim that three- and four-dimensionalism are distinct, rival theories, and they challenge the tacit claim that there are only two metaphysical packages of persistence. If these conclusions are right, then we should reject these two claims.

A rejection of these claims significantly alters the landscape of the debate about persistence. For it alters the diversity of the landscape. It suggests that on some occasions we have less diversity than we thought, because what seemed to be distinct theories are in fact not distinct. And it suggests that on other occasions we have greater diversity than we thought, because there are more theories, or packages, of persistence than previously acknowledged. In essence then, the landscape changes because there are different theoretical players on the metaphysical stage.

Once we see this stage with its new players, we can see which issues are genuine issues, and which mere metaphysical misunderstanding. But in addition we can see the broader structure of the theoretical landscape: we can see not only which theories of persistence are consistent—which metaphysical views fit together to form a consistent metaphysical package—but we can see *how* the components of each package are related.

It might, for instance, have seemed that some components of a package will be consistent with almost any other combination of components. We might have thought that the issue of whether simples persist, or are instantaneous, is largely orthogonal to the issue of what composite objects look like at, and across time. So we might have thought that we can settle the former issue about the nature of simples, independently of any views we have about the nature of composite objects. We have seen, however, that this is not so. One's view about simples does not leave untouched the rest of a theory of persistence. So certain packages of views that we assumed to be consistent, may turn out to be inconsistent.

The message here then, is that we should focus more on complete packages than on individual components of a package. For we cannot determine the plausibility of components of a package, independent of the plausibility of the package as a whole. Of course, in some sense we have always known this. We have always known that commitment to any metaphysical view entails a commitment to certain other metaphysical views. So we have always known that we cannot determine the plausibility of single view, independent of knowing the plausibility of the views it entails. But we don't always act as though we know this to be the case: frequently we focus on components of packages with little regard for consideration of the package as a whole. This book serves to remind us that this is a mistake.

But the lesson of this book is rather broader than just a reminder that we should focus on whole packages rather than just on components of packages. The book makes the bolder claim that we should focus on the entire *diversity* of packages, not simply on individual packages or pairs of packages. Focusing on individual packages allows us to see the relations that hold between the components of *that* package. They allow us to see that certain components of that package entail other components, and they allow us to see whether there are any components that are not entailed by any of the other components. But focusing only on individual packages, or pairs of packages, obscures a whole range of other issues about the packages and their components. For instance, it does not allow us to see how components are related *across* packages.

We have discovered that some components are such that a very small difference in that component, can make a very big difference to the overall theory of persistence that the package offers. That is, some components are what we might think of as *highly effective*: alter the component slightly, and the overall package changes radically. Unitary and non-unitary theories offer a radically different picture of the persisting object, yet they differ only with respect to one component: they differ with respect to the component of composition, and in particular, the component of mereology. But that component *itself* differs only slightly in each case: the only difference between the mereological component of each theory is a difference of one axiom—the axiom that allows for fusions-at-times. So we might say that this is a highly effective component: change it only slightly, and the whole package changes from being a unitary theory to a non-unitary theory. Yet it is only once we see the entire theoretical diversity, that we can see that certain components of packages are highly effective in this manner. And it is not just discerning the highly effective nature of some components that requires taking this broader perspective. In addition to talking of components being highly effective, we can talk of components being *resilient*, where a component is resilient if all of the theories that are minimally plausible, share that same component. The existence of synchronic fusions that coincide with persisting objects at times is, I have argued, an example of a resilient component. That claim is controversial; but the point is that without looking at the diversity of whole packages, it would be impossible either to make, or to evaluate such a claim.

It is by examining the diversity of packages that we discover not only how the packages are related, but also how components are related across packages. We discover, for instance, whether components are resilient or highly effective. Focusing entirely on individual packages, or pairs of packages, leaves many questions not only unanswered, but indeed, unasked. For until we see how the diversity of packages are distributed across logical space, there are simply some questions we do not know to ask. Once we see this distribution, however, we are in a position to draw conclusions not just about the plausibility of individual theories, but about sets of theories. We can draw conclusions about sets of theories that share common components. We can see which theories stand or fall together, given a shared commitment to a particular component. We can see which components are mutually exclusive, and hence which commitments rule out other sorts of commitments. We can ask questions about the relationship between certain sets and sub-sets of theories, and certain components of the theories in those sets. For instance, we can see whether there are any components shared by the entire diversity of theories. We can see whether there are any components shared by the sub-set of plausible theories, or any components shared by the sub-set of implausible theories.

In essence, looking at the manner in which the diversity of theories is distributed across logical space, allows us to see deep structural issues at the core of the debate, and to ask fundamental questions about the nature of those theories and their components. The hope is that this book not only changes the terrain of that logical space by changing our views about theoretical diversity, but that in doing so allows us to see the fundamental issues that lie at the heart of a metaphysics of objects.

NOTES

[1] Balashov (2000a).
[2] Hales and Johnson (2003). p 538.
[3] Variations on this argument can be found in Putnam (1967); Saunders (2002) and Rea (1998).
[4] Hinchliff (1996).
[5] Merricks (1999).
[6] Hales and Johnson (2003).
[7] Hales and Johnson (2003). p 533.
[8] Hales and Johnson (2003). p 533.
[9] Putnam (1967).
[10] Balashov (2000c). p 150.
[11] Balashov (2000c). p 151–153.
[12] Yuri Balashov makes the same point in his (2000a).
[13] Balashov (1999).
[14] Smart (1968).
[15] Quine (1960). pp 172, 253.
[16] Hales and Johnson (2003). pp 534–537. In the original example at each end of the train terrorists have bombs timed to go off. In all other respects the example is the same.
[17] Balashov (2000a).
[18] Balashov (2000a).
[19] I owe this point to David Braddon-Mitchell in discussion.
[20] This terminology is from Lewis (1986) footnote 55 p 76.
[21] I use the locution of objects 'existing at regions' as neutral terminology between objects being identical to the regions at which they exist, and occupying the regions at which they exist.
[22] Sider (2001). pp 110–113.

GLOSSARY

Assertibility Mapping A function that maps the sentences of one theory onto the sentences of another theory just when those sentences are assertible under the same possible situations. See chapter 2, p 7.

Co-compilation The relation that holds between two or more terduring non-fusions whenever those non-fusions materially coincide. Specifically, for any two or more terduring non-fusions *O* and *O** that exist at a time *t*, *O* and *O** are co-compiled at *t* just if (i) *O* is compiled by some fusion *F* at *t*, and (ii) *O** is compiled by some fusion *F** at *t*, and (iii) *F* is identical to *F**. See chapter 5, p 131, 143.

Co-composition The relation that holds between two or more perduring objects at a time when those objects materially coincide. Specifically, for any two or more perduring objects O and O* that exist at a time t, O and O* are co-composed at t just if (i) O has some maximal temporal part F at t, and (ii) O* has some maximal temporal part F* at t and (iii) F is identical to F*. See chapter 6, p 166.

Co-constitution The relation that holds between two or more enduring non-fusions whenever those non-fusions materially coincide. Specifically, for any two or more enduring non-fusions *O* and *O** that exist at a time *t*, *O* and *O** are co-constituted at *t* just if (i) *O* is constituted by some fusion *F* at *t*, and (ii) *O** is constituted by some fusion *F** at *t*, and (iii) *F* is identical to *F**. See chapter 4, p 138.

Compilation According to the unitary four-dimensionalist, the relation that holds between a terduring fusion and a terduring non-fusion at a time, such that the fusion entails the existence of the non-fusion at that time. Specifically, a fusion *F* compiles a (terduring) non-fusion *O* at a time *t* just if (i) *F* and *O* exist at *t*, and (ii) the existence of *F* entails the existence of *O* at *t* and (iii) there is no proper part of *F* whose existence entails the existence of *O* at *t*. See chapter 5, p 130.

Constitution The relation that holds between an enduring non-fusion, and the fusions with which it materially coincides at times. Specifically a fusion *F* constitutes an enduring non-fusion *O* at a time *t* just if (i) *F* and *O* exist at *t*, and (ii) the existence of *F* entails the existence of *O* at *t* and (iii) there is no proper part of *F* whose existence entails the existence of *O* at. See chapter 4, p 101, 112.

Correct Translation A translation is correct just if is a practical translation that preserves truth in virtue of the same truth makers. An inter-theoretic translation is a correct one iff the theories in question are metaphysically equivalent.

Diachronic fusion A fusion of particulars that exist at different times.

Endurance *In general*: The manner in which three-dimensional objects persist, by being wholly present whenever they exist, and strictly identical to themselves at every time at which they exist. *In this book:* The manner in which three-dimensional objects persist. O endures iff it exists at multiple times, and is wholly present at each of those times, where an object is 'wholly present' at a time just in case all of its parts S-*simpliciter* are present at that time. See chapter 3, p 81.

Endurantist *In general,* any three-dimensionalist who holds that persisting objects endure. *In this book*, any non-unitary three-dimensionalist (restricted or unrestricted). See chapter 6.

Enduring Non-fusion Parthood S-*simpliciter* The three-dimensionalist's notion of parthood S-*simpliciter* as it pertains to enduring non-fusions. Specifically, an enduring non-fusion *O* has a part *P* S-*simpliciter* at time *t* just if (i) at *t O* is constituted by some fusion *F* that has *P* M-*simpliciter* and (ii) at *t F* has *P* S-*simpliciter*. See chapter 6 p 157.

Enduring Non-unitary Parthood M-simpliciter A notion appealed to by the non-unitary three-dimensionalist, according to which a non-unitary enduring object *O* has a (spatial) part *P* M-*simpliciter* just if there is some time *t* at which *O* is constituted by a fusion *F*, and *F* has P M-*simpliciter*. See chapter 6, p 157.

Four-dimensionalism *In general*: The thesis that persisting objects are temporally extended and hence have four dimensions. *In this book:* the thesis that every possible persisting object is four-dimensional. See chapter 3, p 53.

Fusion Parthood M-*Simpliciter* A fusion F has a part P M-*simpliciter* just if F fuses P.

Fusion Parthood S-*Simpliciter* A fusion F has a part P S-*simpliciter* at a time t just if F has P M-*simpliciter*, and F has P *t*ly at t.

Mereological Universalism *In general*, the thesis that for any arbitrary set of particulars, there is a fusion of the members of that set. *In this book,* the view that for any arbitrary time t and set S of concrete particulars, there exists a fusion of the members of S at t (a synchronic fusion-at-a-time) and for any arbitrary set S^* of synchronic fusions that exist at distinct times $t_1 \ldots t_n$, there exists a diachronic fusion of the members of S^*. See chapter 4, p 114.

Metaphysical Equivalence Any two theories are metaphysically equivalent just if they have the same truth makers. See also correct intertranslatability. See chapter 2.

Non-mereological Universalism The thesis that any combination of particulars at and across time composes some individual. Specifically, the view that for any arbitrary time t and set S of concrete particulars, there exists a fusion of the members of S at t (a synchronic fusion-at-a-time) and for any arbitrary set S^* of synchronic fusions that exist at distinct times $t_1 \ldots t_n$, there exists an enduring object O that is at each of those times, constituted by one of those fusions. See chapter 4, p 115.

Non-Unitary Thesis The thesis that for every persisting object O, and every time t during which O exists, there exists some instantaneous object O^* that overlaps or coincides with O at t.

Non-unitary thesis (3D) The thesis that for every persisting object O and time t at which O exists, there exists some synchronic fusion S such that O is constituted by S at t. See chapter 4, p 113.

Non-unitary thesis (4D) The thesis that for every persisting object O and time t at which O exists, there exists some synchronic fusion S, such that S is an instantaneous temporal part of O at t. See chapter 4, p 113.

Perdurance An object O perdures iff it is a mereological fusion of temporal parts. See chapter 3, p 55.

Perdurantism The thesis that every possible persisting object perdures. See chapter 3, p 56.

Perduring Parthood M-*simpliciter* A perduring object O has a (spatial) part P M-*simpliciter* just if there is some time t at which O has as a temporal part synchronic fusion F, and F has P M-*simpliciter*. See chapter 6, p 158.

Perduring Parthood S-*simpliciter* A perduring object O has (spatial) part P S-*simpliciter* at t just if at t, O has as a temporal part some synchronic fusion F that exists at t, and F has P M-*simpliciter*. See chapter 6, p 158.

Persisting fusion A persisting fusion composed of persisting simples, whether these are enduring simples or terduring simples.

Persisting non-fusions A persisting object that has parts at times, but which is not a fusion of persisting simples; instead, it is related in certain ways to fusions (persisting or synchronic) at times.

Practical Translation An assertibility mapping that is truth preserving. See chapter 2, p 8, 9–17.

Restricted Non-unitary Four-dimensionalism Also known as **restricted perdurantism**. This view combines restricted composition with perdurantism (non-unitary four-dimensionalism). Specifically, it is the view that for some times $t \dots t_n$ and some sets $S \dots S_n$ of concrete particulars, there exist synchronic fusions $F \dots F_n$ of the members of those sets at those times and (ii) for some synchronic fusions $F \dots F_n$ that exist at distinct times $t_1 \dots t_n$, there exist diachronic fusions of those fusions. See chapter 6, p 155.

Restricted Non-unitary Three-dimensionalism Also known as **restricted endurantism**. The view that combines a non-unitary three-dimensionalism with restricted composition. Specifically, the view that: for some times $t \dots t_n$ and some sets $S \dots S_n$ of concrete particulars, there exist synchronic fusions $F \dots F_n$ of the members of those sets at those times and (ii) for some synchronic fusions $F \dots F_n$ that exist at distinct times $t_1 \dots t_n$, there exist enduring objects $O \dots O_n$ that are constituted at each of those times, by those fusions. See chapter 6, p 107.

Restricted Unitary Four-dimensionalism The view that combines the unitary four-dimensionalism with the thesis that composition is restricted. Specifically, the view that (i) for some sets $S....S_n$ of temporally overlapping terduring simples, there exist the fusions $F \dots F_n$ of the members of those sets and (ii) for some fusions $F \dots F_n$ and times $t \dots t_n$ at which those fusions exist, there exist terduring non-fusions $O....O_n$ which are compiled by those fusions at those times. See chapter 5, p 132.

Restricted Unitary Three-dimensionalism The view that combines unitary three-dimensionalism with the thesis that composition is restricted. Specifically, the thesis that for some sets $S....S_n$ of temporally overlapping simples, there exist fusions $F \dots F_n$ of the members of those sets and (ii) for some fusions $F \dots F_n$ and times $t \dots t_n$ at which those fusions exist, there exist enduring non-fusions $O....O_n$ which are constituted by those fusions at those times. See chapter 5, p 107.

Synchronic fusion An instantaneous object that is either (i) the fusion of instantaneous particulars, all of which exist at the same time, or (ii) the fusion, at-a-time, of particulars which may exist at other times.

Strongly gain or lose parts A persisting object O strongly loses or gains a part P iff P is part of O at some time t, O exists at t^*, P exists at t^*, and P is not part of O at t^*. See chapter 4, p 98–99.

Terdurance The manner of persistence of unitary four-dimensional objects. An object O terdures just if it exists at multiple temporal locations in virtue of being a temporally extended temporal simple that does not wholly exist at any of the temporal locations at which it exists. See chapter 3, p 56.

Terdurantism The thesis that every possible persisting object terdures. See chapter 3, p 57.

Terduring Non-fusion Parthood S-*simpliciter* The terdurantist notion of parthood S-*simpliciter* as it pertains to terduring non-fusions. Specifically, a terduring non-fusion O has a part P S-*simpliciter* at a time t just if (i) at t O is compiled by some fusion F that has P M-*simpliciter* and (ii) at t F has P S-*simpliciter*. Chapter 5, p 136.

Three-dimensionalism *In general:* The thesis that persisting objects are three-dimensional and persist by enduring. *In this book:* The thesis that all possible persisting objects are three dimensional. See chapter 3, p 60.

Unitary Theories The view that there do not exist instantaneous or short-lived objects that coincide with persisting objects whenever those persisting objects exist.

Unrestricted Non-unitary Four-dimensionalism Also known as **unrestricted perdurantism**. The combination of perdurantism and mereological universalism. The thesis that for any arbitrary time t and set S of concrete particulars there exists a fusion of the members of S at t (a synchronic fusion-at-a-time) and for any arbitrary set S^* of synchronic fusions that exist at distinct times $t_1 \ldots t_n$, there exists a diachronic fusion of the members of S^*. See Chapter 6, p 154.

Unrestricted Non-unitary Three-dimensionalism Also known as **unrestricted endurantism.** A three-dimensionalist view that embraces non-mereological universalism. The thesis that for any arbitrary time t and set S of concrete particulars there exists a fusion of the members of S at t (a synchronic fusion-at-a-time) and for any arbitrary set S^* of synchronic fusions that exist at distinct times $t_1 \ldots t_n$, there exists an enduring object O that is at each of those times, constituted by one of those fusions. See chapter 4 p 32 and chapter 6 p 154.

Unrestricted Unitary Four-dimensionalism The thesis that persisting objects are four-dimensional temporal simples, combined with the view that composition unrestricted, understood in non-mereological terms. Specifically, the view that (i) for any arbitrary set S of temporally overlapping terduring simples, there exists a fusion of the members of S and (ii) for any arbitrary set S of fusions, and any arbitrary times at which those fusions exist, there exists some terduring non-fusion that is compiled by those fusions at those times. See chapter 5, p 132.

Unrestricted Unitary Three-dimensionalism The thesis that combines unitary three-dimensionalism with the view that composition is unrestricted. Specifically, the view that (i) for any arbitrary set S of temporally overlapping simples, there exists a fusion of the members of S and (ii) for any arbitrary set S of fusions, and any arbitrary times at which those fusions exist, there exists some enduring non-fusion that is constituted by those fusions at those times. See chapter 5, p 107.

Weakly gain Or Lose Parts A persisting object O weakly gains or loses a part P (i) P is part of O at all times at which P exists and (ii) there is some time t at which O exists, and P does not. See chapter 4, p 98–99.

BIBLIOGRAPHY

Armstrong, D. (1980). 'Identity through time' pp 67–68 in *Time and Cause* van Inwagen (ed). Dordrecht: Reidel.

Baker, L.R. (2000). *Persons and Bodies*. Cambridge University Press.

Baker, L.R. (1997). "Why Constitution is not Identity." *Journal of Philosophy* 94: 599–621.

Balaguer, M. (1998). *Platonism and Anti-Platonism in Mathematics*. New York: Oxford University Press.

Balashov, Y. (1999). "Relativistic Objects." *Nous*. 33(4): 644–662.

Balashov, Y. (2000a). "Persistence and Space-time: Philosophical Lessons of the Pole and Barn." *The Monist*. 83(3): 321–240.

Balashov, Y. (2000b). "Relativity and Persistence." *Philosophy of Science*. 67(3): 549–562.

Balashov, Y. (2000c). "Enduring and Perduring Objects in Minkowski Space-Time." *Philosophical Studies* 99: 129–166.

Balashov, Y. (2002). "On Stages, Worms and Relativity." *Philosophy* 50: 223–252.

Balashov, Y. (2003). "Restricted Diachronic Composition, Immanent Causality, and Objecthood: A Reply to Hudson" *Philosophical Papers* 32(1) 23–30.

Benacerraf, P. (1973). "Mathematical Truth" *Journal of Philosophy* 70: 661–80.

Braddon-Mitchell, D. (2004). "Masters of our Meanings" *Philosophical Studies*; 118(1–2): 133–152.

Braddon-Mitchell, D. (2003). 'Qualia and analytical conditionals' *Journal of Philosophy* 100: 111–135.

Braddon-Mitchell, D. and Jackson, F. (1996). *Philosophy of Mind and Cognition*. Oxford University Press.

Brogaard, B. (2000). "Presentist Four Dimensionalism." *The Monist* 83(3): 341–356.

Chalmers, D. (2004). "Epistemic Two Dimensional Semantics" *Philosophical Studies*; 118(1–2): 153–226.

Chambers, T. (1999). "Time travel: How not to Defuse the Principal Paradox." *Ratio* 12(3): 296–301.

Davidson, D. (1973). 'Radical Interpretation' in *Inquiries into Truth and Interpretation* Oxford: OUP.

DeVito, S. (1997). "A gruesome problem for the curve-fitting solution." *British Journal for the Philosophy of Science* 48(3): 391–396.

Doepke, F. (1982). "Spatially Coinciding Objects". *Ratio* 24: 45–60.

Dowe, P. and Barker, S. (2003). 'Paradoxes of multi-location' *Analysis* 63(2) 106–114.

Evans, G. (1978). Can There Be Vague Objects? *Analysis* 38: 208.

Gallois, A. (1998). *Occasions of identity: a study in the metaphysics of persistence, change and sameness.* Oxford: Clarendon Press.
Gibbard, A. (1975). "Contingent Identity." *Journal of Philosophical Logic* 4: 197–221.
Grey, W. (1999). "Troubles with Time Travel". *Philosophy* 74: 55–70.
Griffin, N. (1978). "Relative Identity Relations." *Mind* 87: 576–581.
Hales, S.D and Johnson, T.A. (2003). "Endurantism, Perdurantism and Special Relativity" *The Philosophical Quarterly* 53: 213 524–539.
Haslanger, S. (1989). "Endurance and Temporary Intrinsics. *Analysis* 49: 119–125.
Heller, M. (1990). *The Ontology of Physical Objects: Four Dimensional Hunks of Matter.* Cambridge: Cambridge University Press.
Heller, M. (1984). "Temporal Parts of Four-Dimensional Objects." *Philosophical Studies.* 46: 323–334.
Hinchliff, M. (1996). "The puzzle of change." *Philosophical Perspectives 10, Metaphysics* 119–133.
Hirsch, E. (2002). "Quantifier Variance and Realism" *Philosophical Issues, 12 Realism and Relativism* 51–73.
Hoefer, C. and Rosenberg, A. (1994). "Empirical Equivalence, Underdetermination and Systems of the World." *Philosophy of Science.* 61: 592–607.
Horwich, P. (1975) "On Some Alleged Paradoxes of Time Travel." *Journal of Philosophy* 72: 432–444.
Howard-Snyder, F. (1991). "*De Re* Modality entails *De Re* Vagueness. *Philosophical Quarterly* 72: 101–112.
Hyde, D. (1998). "Vagueness, Ontology and Supervenience." *The Monist.* 81: 297–312.
Jackson, F. (2004). "Why we need A-intensions." *Philosophical Studies.* 118(1–2): 257–277.
Johnston, M. (1992). "Constitution Is Not Identity". *Mind* 101: 89–105.
Johnston, M. (1987). "Is There a Problem about Persistence?" *The Aristotelian Society* Supp 61: 107–135.
Koslicki, K. (2003). "The crooked path from vagueness to four dimensionalism." *Philosophical Studies*, 114: 107–134.
Kriple, S. (1972). *Naming and Necessity.* Cambridge: Harvard University Press.
Leibniz, G.W. (1696). "On the Principle of Indiscernibles." in G.H.R Parkinson ed. *Philosophical Writings* 1995, Everyman.
Lewis, D. (1991). *Parts of Classes.* Oxford: Blackwell.
Lewis, D. (1988). "Vague Identity: Evans Misunderstood." *Analysis* 48: 128–30.
Lewis, D. (1986). *On the Plurality of Worlds.* New York Blackwell Press.
Lewis, D. (1983). "Survival and Identity." in *Philosophical Papers Vol I* Oxford University Press pp 55–77.
Lewis, D. (1976). "The Paradoxes of Time Travel" *American Philosophical Quarterly* 13: 145–152.
Lewis, D. (1972). "Psychophysical and Theoretical Identifications." *Australasian Journal of Philosophy.* 3: 249–258.
Loar, B. (1981). *Mind and Meaning.* Cambridge University Press.
Lycan, W. (199). *Mind and Cognition.* Oxford: Blackwell Publishers.
Markosian, N. (1994). "The 3D/4D Controvery and Non-Present Objects." *Philosphical Papers* 23: 243–49.
Markosian, N. (1998). 'Simples' The Australasian Journal of Philosophy. 76(2) 213–229.
Markosian, N. (2004). 'Simples, stuff and simple people.' *The Monist* 87(3): 405–421.

McCall, S. (1994). *A Model of the Universe: Space-Time, Probability, and Decision.* Oxford: Clarendon Press.

Merricks, T. (1994). "Endurance and Indiscernibility." *The Journal of Philosophy.* 91: 165–84.

Merricks, T. (1995). "On the Incompatibility of Enduring and Perduring Entities." *Mind* 104: 523–41.

Merricks, T. (1999). "Persistence, Parts and Presentism." *Nous* 33(3): 421–438.

Merricks, T. (2000). "No Statues." *Australasian Journal of Philosophy*, 78(8): 47–52.

Merricks, T. (2001). *Objects and Persons.* Oxford University Press.

Myro, G. (1986). "Identity and Time". in *The Philosophical Grounds of Rationality* ed R. Grandy and R. Warner. New York Clarendon Press.

Noonan, H. (1993). "Constitution is identity" *Mind* 102: 133–146.

Nozick, R. (1982). *Philosophical Explanations*, MA: Harvard University Press.

Parfit, D. (1984). *Reasons and Persons.* Oxford: Oxford University Press.

Parsons, J. (2000). "Must a Four Dimensionalist Believe in Temporal Parts?" *The Monist* 83(3): 399–418.

Potemkin, I.I. (2004). 'Overcharging of complexes formed by charged interpenetrable macroins'. *Europhysics Letters* 68 (4) 487–494.

Putnam, H. (1988). "After Metaphysics, What?" in *Metaphysics: The Big Questions.* eds D. Zimmerman and P. van Inwagen. Oxford: Basil Blackwell Press. pp 388–392.

Putnam, H. (1987). "Truth and Convention." in *Metaphysics: The Big Questions.* eds D. Zimmerman and P. van Inwagen. Oxford: Basil Blackwell Press. pp 392–398.

Quine, W.V.O. (1960). *Word and Object.* Cambridge: MIT Press.

Quine, W.V.O. (1975). "On Empirically Equivalent Systems in the World." *Erkenntnis* 9: 313–328.

Robinson, D. (1985). "Can Amoebae Divide without Multiplying?" *Australasian Journal of Philosophy* 63: 299–319.

Robinson, D. (1982). "Reidentifying Matter". *Philosophical Review* 91: 317–342.

Sainsbury, R.M. (1989). 'What is a Vague Object.' *Analysis* 49: 99–103.

Shoemaker, S. (1979). 'Identity, properties and causation.' *Midwest Studies in Philosophy.*

Sidelle, A. (2002). "Is there a True Metaphysics of Material Objects?" *Philosophical Issues 12 Realism and Relativism* 118–145.

Sider, T. (2003). "Against Vague Existence." *Philosophical Studies*, 114: 135–146.

Sider, T. (2002). "Time Travel, Coincidences and Counterfactuals." *Philosophical Studies* 110: 115–138.

Sider, T. (2001). *Four-dimensionalism: an ontology of persistence and time.* Oxford University Press.

Sider, T. (2000). "The Stage View and Temporary Intrinsics." *Analysis* 60: 84–88.

Sider, T. (1999). "Presentism and Ontological Commitment" *Journal of Philosophy* 94(7) 325–347.

Simons, P. (1987). *Parts: A Study in Ontology.* Oxford: Clarendon Press.

Simons, P. (1985). "Coincidence of Things of a Kind." *Mind* 94: 70–75.

Smart, B. (1973). "The Ship of Theseus, The Parthenon and Disassembled Objects." *Analysis* 33(4): 299–319.

Smart, B. (1972). "How to Reidentify the Ship of Theseus." *Analysis* 32(3): 145–158.

Smart, J.J.C. (1968). *Between Philosophy and Science: An Introduction to the Philosophy of Science.* Random House: New York.

Sober, E. (1979). *Simplicity.* Oxford: Clarendon Press.

Stone, J. (2003). "On staying the same" *Analysis* 63(4): 288–292.

Thomson, J.J. (1998) The Statue and the Clay. *Nous* 32: 149–173.

Thomson, J.J. (1983). "Parthood and Identity Across Time." *Journal of Philosophy* 80: 201–220.

Tye, M. (1990). 'Vague Objects.' *Mind* 9: 535–557.

Van Inwagen, P. (1987). "When are Objects Parts?" *Philosophical Perspectives 1. Metaphysics:* 21–47.

Van Inwagen, P. (1990). *Material Beings.* Cornell University Press.

Van Inwagen, P. (1990a). 'Four-Dimensional Objects." *Nous* 24: 245–255.

Van Inwagen, P. (2002). "The Number of Things" *Philosophical Issues 12 Realism and Relativism* 176–196.

Wiggins, D. (1980). *Sameness and Substance.* Oxford: Blackwell Press.

Wiggins, D. (1968). "On Being in the Same Place at the Same Time." *Philosophical Review* 77: 90–95.

Zalta, E. and Colyvan, M. (1999). "Mathematics: Truth and Fiction?" *Philosophia Mathematica* 7(3): 336–349.

INDEX

PHILOSOPHICAL STUDIES SERIES

1. Jay F. Rosenberg: *Linguistic Representation.* 1974 ISBN 90-277-0533-X
2. Wilfrid Sellars: *Essays in Philosophy and Its History.* 1974 ISBN 90-277-0526-7
3. Dickinson S. Miller: *Philosophical Analysis and Human Welfare.* Selected Essays and Chapters from Six Decades. Edited with an Introduction by Lloyd D. Easton. 1975 ISBN 90-277-0566-6
4. Keith Lehrer (ed.): *Analysis and Metaphysics.* Essays in Honor of R. M Chisholm. 1975 ISBN 90-277-0571-2
5. Carl Ginet: *Knowledge, Perception, and Memory.* 1975 ISBN 90-277-0574-7
6. Peter H. Hare and Edward H. Madden: *Causing, Perceiving and Believing.* An Examination of the Philosophy of C. J. Ducasse.1975 ISBN 90-277-0563-1
7. Hector-Neri Castañeda: *Thinking and Doing.* The Philosophical Foundations of Institutions. 1975 ISBN 90-277-0610-7
8. John L. Pollock: *Subjunctive Reasoning.* 1976 ISBN 90-277-0701-4
9. Bruce Aune: *Reason and Action.* 1977 ISBN 90-277-0805-3
10. George Schlesinger: *Religion and Scientific Method.* 1977 ISBN 90-277-0815-0
11. Yirmiahu Yovel (ed.): *Philosophy of History and Action.* Papers presented at the First Jerusalem Philosophical Encounter (December 1974). 1978 ISBN 90-277-0890-8
12. Joseph C. Pitt (ed.): *The Philosophy of Wilfrid Sellars: Queries and Extensions.* 1978 ISBN 90-277-0903-3
13. Alvin I. Goldman and Jaegwon Kim (eds.): *Values and Morals.* Essays in Honor of William Frankena, Charles Stevenson, and Richard Brandt. 1978 ISBN 90-277-0914-9
14. Michael J. Loux: *Substance and Attribute.* A Study in Ontology. 1978 ISBN 90-277-0926-2
15. Ernest Sosa (ed.): *The Philosophy of Nicholas Rescher.* Discussion and Replies. 1979 ISBN 90-277-0962-9
16. Jeffrie G. Murphy: *Retribution, Justice, and Therapy.* Essays in the Philosophy of Law. 1979 ISBN 90-277-0998-X
17. George S. Pappas (ed.): *Justification and Knowledge.* New Studies in Epistemology. 1979 ISBN 90-277-1023-6
18. James W. Cornman: *Skepticism, Justification, and Explanation.* With a Bibliographic Essay by Walter N. Gregory. 1980 ISBN 90-277-1041-4
19. Peter van Inwagen (ed.): *Time and Cause.* Essays presented to Richard Taylor. 1980 ISBN 90-277-1048-1
20. Donald Nute: *Topics in Conditional Logic.* 1980 ISBN 90-277-1049-X
21. Risto Hilpinen (ed.): *Rationality in Science.* Studies in the Foundations of Science and Ethics. 1980 ISBN 90-277-1112-7
22. Georges Dicker: *Perceptual Knowledge.* An Analytical and Historical Study. 1980 ISBN 90-277-1130-5
23. Jay F. Rosenberg: *One World and Our Knowledge of It.* The Problematic of Realismin Post-Kantian Perspective. 1980 ISBN 90-277-1136-4
24. Keith Lehrer and Carl Wagner: *Rational Consensus in Science and Society.* A Philosophical and Mathematical Study. 1981 ISBN 90-277-1306-5
25. David O'Connor: *The Metaphysics of G. E. Moore.* 1982 ISBN 90-277-1352-9

PHILOSOPHICAL STUDIES SERIES

26. John D. Hodson: *The Ethics of Legal Coercion.* 1983 ISBN 90-277-1494-0
27. Robert J. Richman: *God, Free Will, and Morality.* Prolegomena to a Theory of Practical Reasoning. 1983 ISBN 90-277-1548-3
28. Terence Penelhum: *God and Skepticism.* A Study in Skepticism and Fideism. 1983 ISBN 90-277-1550-5
29. James Bogen and James E. McGuire (eds.): *How Things Are.* Studies in Predication and the History of Philosophy of Science. 1985 ISBN 90-277-1583-1
30. Clement Dore: *Theism.* 1984 ISBN 90-277-1683-8
31. Thomas L. Carson: *The Status of Morality.* 1984 ISBN 90-277-1619-9
32. Michael J. White: *Agency and Integrality.* Philosophical Themes in the Ancient Discussions of Determinism and Responsibility. 1985 ISBN 90-277-1968-3
33. Donald F. Gustafson: *Intention and Agency.* 1986 ISBN 90-277-2009-6
34. Paul K. Moser: *Empirical Justification.* 1985 ISBN 90-277-2041-X
35. Fred Feldman: *Doing the Best We Can.* An Essay in Informal Deontic Logic. 1986 ISBN 90-277-2164-5
36. G. W. Fitch: *Naming and Believing.* 1987 ISBN 90-277-2349-4
37. Terry Penner: *The Ascent from Nominalism.* Some Existence Arguments in Plato's Middle Dialogues. 1987 ISBN 90-277-2427-X
38. Robert G. Meyers: *The Likelihood of Knowledge.* 1988 ISBN 90-277-2671-X
39. David F. Austin (ed.): *Philosophical Analysis.* A Defense by Example. 1988 ISBN 90-277-2674-4
40. Stuart Silvers (ed.): *Rerepresentation.* Essays in the Philosophy of Mental Representation. 1988 ISBN 0-7923-0045-9
41. Michael P. Levine: *Hume and the Problem of Miracles.* A Solution. 1989 ISBN 0-7923-0043-2
42. Melvin Dalgarno and Eric Matthews (eds.): *The Philosophy of Thomas Reid.* 1989 ISBN 0-7923-0190-0
43. Kenneth R. Westphal: *Hegel's Epistemological Realism.* A Study of the Aim and Method of Hegel's *Phenomenology of Spirit.* 1989 ISBN 0-7923-0193-5
44. John W. Bender (ed.): *The Current State of the Coherence Theory.* Critical Essays on the Epistemic Theories of Keith Lehrer and Laurence BonJour, with Replies. 1989 ISBN 0-7923-0220-6
45. Roger D. Gallie: *Thomas Reid and 'The Way of Ideas'.* 1989 ISBN 0-7923-0390-3
46. J-C. Smith (ed.): *Historical Foundations of Cognitive Science.* 1990 ISBN 0-7923-0451-9
47. John Heil (ed.): *Cause, Mind, and Reality.* Essays Honoring C. B. Martin. 1989 ISBN 0-7923-0462-4
48. Michael D.Roth and Glenn Ross(eds.):*Doubting.* Contemporary Perspectives on Skepticism. 1990 ISBN 0-7923-0576-0
49. Rod Bertolet: *What is Said.* A Theory of Indirect Speech Reports. 1990 ISBN 0-7923-0792-5
50. Bruce Russell (ed.): *Freedom, Rights and Pornography.* A Collection of Papers by Fred R. Berger. 1991 ISBN 0-7923-1034-9
51. Kevin Mulligan (ed.): *Language, Truth and Ontology.* 1992 ISBN 0-7923-1509-X

PHILOSOPHICAL STUDIES SERIES

52. Jesús Ezquerro and Jesús M. Larrazabal (eds.): *Cognition, Semantics and Philosophy*. Proceedings of the First International Colloquium on Cognitive Science. 1992 ISBN 0-7923-1538-3
53. O.H. Green: *The Emotions*. A Philosophical Theory. 1992 ISBN 0-7923-1549-9
54. Jeffrie G. Murphy: *Retribution Reconsidered*. More Essays in the Philosophy of Law. 1992 ISBN 0-7923-1815-3
55. Phillip Montague: *In the Interests of Others*. An Essay in Moral Philosophy. 1992 ISBN 0-7923-1856-0
56. Jacques-Paul Dubucs (ed.): *Philosophy of Probability*. 1993 ISBN 0-7923-2385-8
57. Gary S. Rosenkrantz: *Haecceity*. An Ontological Essay. 1993 ISBN 0-7923-2438-2
58. Charles Landesman: *The Eye and the Mind*. Reflections on Perception and the Problem of Knowledge. 1994 ISBN 0-7923-2586-9
59. Paul Weingartner (ed.): *Scientific and Religious Belief*. 1994 ISBN 0-7923-2595-8
60. Michaelis Michael and John O'Leary-Hawthorne (eds.): *Philosophy in Mind*. The Place of Philosophy in the Study of Mind. 1994 ISBN 0-7923-3143-5
61. William H. Shaw: *Moore on Right and Wrong*. The Normative Ethics of G.E. Moore. 1995 ISBN 0-7923-3223-7
62. T.A. Blackson: *Inquiry, Forms, and Substances*. A Study in Plato's Metaphysics and Epistemology. 1995 ISBN 0-7923-3275-X
63. Debra Nails: *Agora, Academy, and the Conduct of Philosophy*. 1995 ISBN 0-7923-3543-0
64. Warren Shibles: *Emotion in Aesthetics*. 1995 ISBN 0-7923-3618-6
65. John Biro and Petr Kotatko (eds.): *Frege: Sense and Reference One Hundred Years Later. 1995* ISBN 0-7923-3795-6
66. Mary Gore Forrester: *Persons, Animals, and Fetuses*. An Essay in Practical Ethics. 1996 ISBN 0-7923-3918-5
67. K. Lehrer, B.J. Lum, B.A. Slichta and N.D. Smith (eds.): *Knowledge, Teaching and Wisdom*. 1996 ISBN 0-7923-3980-0
68. Herbert Granger: *Aristotle's Idea of the Soul*. 1996 ISBN 0-7923-4033-7
69. Andy Clark, Jesús Ezquerro and Jesús M. Larrazabal (eds.): *Philosophy and Cognitive Science: Categories, Consciousness, and Reasoning*. Proceedings of the Second International Colloquium on Cogitive Science. 1996 ISBN 0-7923-4068-X
70. J. Mendola: *Human Thought*. 1997 ISBN 0-7923-4401-4
71. J. Wright: *Realism and Explanatory Priority*. 1997 ISBN 0-7923-4484-7
72. X. Arrazola, K. Korta and F.J. Pelletier (eds.): *Discourse, Interaction and Communicationion*. Proceedings of the Fourth International Colloquium on Cognitive Science. 1998 ISBN 0-7923-4952-0
73. E. Morscher, O. Neumaier and P. Simons (eds.): *Applied Ethics in a Troubled World*. 1998 ISBN 0-7923-4965-2
74. R.O. Savage: *Real Alternatives, Leibniz's Metaphysics of Choice*. 1998 ISBN 0-7923-5057-X
75. Q. Gibson: *The Existence Principle*. 1998 ISBN 0-7923-5188-6
76. F. Orilia and W.J. Rapaport (eds.): *Thought, Language, and Ontology*. 1998 ISBN 0-7923-5197-5

PHILOSOPHICAL STUDIES SERIES

77. J. Bransen and S.E. Cuypers (eds.): *Human Action, Deliberation and Causation.* 1998 ISBN 0-7923-5204-1
78. R.D. Gallie: *Thomas Reid: Ethics, Aesthetics and the Anatomy of the Self* . 1998 ISBN 0-7923-5241-6
79. K. Korta, E. Sosa and X. Arrazola (eds.): *Cognition, Agency and Rationality.* Proceedings of the Fifth International Colloquium on Cognitive Science. 1999 ISBN 0-7923-5973-9
80. M. Paul: *Success in Referential Communication.* 1999 ISBN 0-7923-5974-7
81. E. Fischer: *Linguistic Creativity.* Exercises in 'Philosophical Therapy'. 2000 ISBN 0-7923-6124-5
82. R. Tuomela: *Cooperation.* A Philosophical Study. 2000 ISBN 0-7923-6201-2
83. P. Engel (ed.): *Believing and Accepting.* 2000 ISBN 0-7923-6238-1
84. W.L. Craig: *Time and the Metaphysics of Relativity.* 2000 ISBN 0-7923-6668-9
85. D.A. Habibi: *John Stuart Mill and the Ethic of Human Growth.* 2001 ISBN 0-7923-6854-1
86. M. Slors: *The Diachronic Mind.* An Essay on Personal Identity, Psychological Continuity and the Mind-Body Problem. 2001 ISBN 0-7923-6978-5
87. L.N. Oaklander (ed.): *The Importance of Time.* Proceedings of the Philosophy of Time Society, 1995–2000. 2001 ISBN 1-4020-0062-6
88. M. Watkins: *Rediscovering Colors.* A Study in Pollyanna Realism. 2002 ISBN 1-4020-0737-X
89. W.F. Vallicella: *A Paradigm Theory of Existence.* Onto–Theology Vindicated. 2002 ISBN 1-4020-0887-2
90. M. Hulswit: *From Cause to Causation.* A Peircean Perspective. 2002 ISBN 1-4020-0976-3; Pb 1-4020-0977-1
91. D. Jacquette (ed.): *Philosophy, Psychology, and Psychologism.* Critical and Historical Readings on the Psychological Turn in Philosophy. 2003 ISBN 1-4020-1337-X
92. G. Preyer, G. Peterand, M. Ulkan (eds.): *Concepts of Meaning.* Framing an Integrated Theory of Linguistic Behavior. 2003 ISBN 1-4020-1329-9
93. W. de Muijnck: *Dependencies, Connections, and Other Relations.* A Theory of Mental Causation. 2003 ISBN 1-4020-1391-4
94. N. Milkov: *A Hundred Years of English Philosophy.* 2003 ISBN 1-4020-1432-5
95. E.J. Olsson (ed.): *The Epistomology of Keith Lehrer.* 2003 ISBN 1-4020-1605-0
96. D.S. Clarke: *Sign Levels.* Language and Its Evolutionary Antecedents. 2003 ISBN 1-4020-1650-6
97. A. Meirav: *Wholes, Sums and Unities.* 2003 ISBN 1-4020-1660-3
98. C.H. Conn: *Locke on Esence and Identity.* 2003 ISBN 1-4020-1670-0
99. J.M. Larrazabal and L.A. Pérez Miranda (eds.): *Language, Knowledge, and Representation.* Proceedings of the Sixth International Colloquium on Cognitive Science (ICCS-99). 2004 ISBN 1-4020-2057-0
100. P. Ziff: *Moralities.* A Diachronic Evolution Approach. 2004 ISBN 1-4020-1891-6
101. J.A. Corlett: Terrorism: *A Philosophical Analysis. 2003* ISBN 1-4020-1694-8; Pb 1-4020-1695-6

PHILOSOPHICAL STUDIES SERIES

102. K. Korta and J.M. Larrazabal (eds.): *Truth, Rationality, Cognition, and Music*. Proceedings of the Seventh International Colloquium on Cognitive Science. 2004 ISBN 1-4020-1912-2
103. T.M. Crisp, M. Davidson and D. Vander Laan (eds.): Knowledge and Reality. 2006 ISBN 1-4020-4732-0
104. S.E. Boër (ed.): *Thought-Contents*. On the Ontology of Belief and the Semantics of Belief Attribution. 2006 ISBN 1-4020-5084-4
105. A. Voltolini (ed.): *How Ficta Follow Fiction*. A Syncretistic Account of Fictional Entities. 2006 ISBN 1-4020-5146-8
106. K.L. Miller (ed.): *Issues in Theoretical Diversity*. Persistence, Composition, and Time. 2006 ISBN 1-4020-5255-3
107. M. Kieran (eds.): *Knowing Art*. Essays in Aesthetics and Epistemology. 2006 ISBN 1-4020-5264-2

9 781402 052552

MIX
Papier aus verantwortungsvollen Quellen
Paper from responsible sources
FSC® C105338

If you have any concerns about our products,
you can contact us on
ProductSafety@springernature.com

In case Publisher is established outside the EU,
the EU authorized representative is:
Springer Nature Customer Service Center GmbH
Europaplatz 3, 69115 Heidelberg, Germany

Printed by Libri Plureos GmbH
in Hamburg, Germany